AF882123

CHESTER THROUGH DERRY CONODONTS AND STRATIGRAPHY OF NORTHERN CLARK AND SOUTHERN LINCOLN COUNTIES, NEVADA

CHESTER THROUGH DERRY CONODONTS AND STRATIGRAPHY OF NORTHERN CLARK AND SOUTHERN LINCOLN COUNTIES, NEVADA

BY
G. D. WEBSTER

UNIVERSITY OF CALIFORNIA PRESS
BERKELEY AND LOS ANGELES
1969

University of California Publications in Geological Sciences

Advisory Editors: W. A. Clemens, G. H. Curtis, A. E. J. Engel, C. G. Higgins, R. M. Kleinpell, Jean Langenheim, M. A. Murphy, R. L. Shreve, C. A. Wahrhaftig, A. C. Waters, G. W. Wetherill

Volume 79
Approved for publication April 19, 1968
Issued April 11, 1969
Price, $4.00

University of California Press
Berkeley and Los Angeles
California

University of California Press, Ltd.
London, England

Copyright © 1969 by the Regents of the University of California
Printed in the United States of America

CONTENTS

CHESTER THROUGH DERRY CONODONTS AND STRATIGRAPHY OF NORTHERN CLARK AND SOUTHERN LINCOLN COUNTIES, NEVADA

BY

G. D. WEBSTER

ABSTRACT

A BIOSTRATIGRAPHIC STUDY in northern Clark and southern Lincoln counties, Nevada, has revealed the presence of a remarkably complete, fossiliferous section of Late Mississippian through Early Pennsylvanian age. Parts or all of four formations, the Battleship Wash, the Indian Springs, the Bird Spring, and the Chainman shale, along with their fossil faunas, were studied at eleven localities concentrated in the Las Vegas, Arrow Canyon, Meadow Valley, and Delamar ranges. Although few lithologic units are continuous, the Indian Springs and the overlying Bird Spring are recognizable throughout the area and each is subdivided into four informal units. The Battleship Wash formation which occurs in the south is at least partly equivalent to the Chainman shale in the north.

Three new species of conodonts, *Spathognathodus rexroadi, S. lanceolatus,* and *Neoprioniodus? expandofundus,* are described and shown to be confined to useful zones. The conodonts and associated megafossils indicate that the upper part of the Indian Springs formation and the basal part of the Bird Spring formation are referable to the European *Homoceras* Zone. The Mississippian-Pennsylvanian boundary occurs within the basal part of the Bird Spring formation and coincides with the upper limit of the *Rhipidomella nevadensis* Assemblage Zone.

Conodonts are abundant throughout the section and two faunizones are proposed, the *Gnathodus girtyi simplex* Zone of middle and late Chester age and the *Streptognathodus noduliferus-Idiognathoides convexus* Zone of early and middle Morrow age.

The fusuline zones of *Millerella* and *Profusulinella* are recognized and thought to be equivalents of these zones in the midcontinent. Because of the abundance of *Chaetetes* a faunizone of *Fusulinella-Chaetetes* is proposed for rocks of late Derry age. This zone is recognized only in southern Nevada.

INTRODUCTION

THIS STUDY was undertaken for the purpose of establishing conodont faunal zones, correlating measured sections, and describing the conodont taxa of Chester through Derry age rocks in southern Nevada. Eleven stratigraphic sections were designated A through K (fig. 1) and given names derived from nearby geographic features. Each section was measured, described, and sampled for fossils during the spring of 1963 (see Appendix for descriptions of measured sections). Selection was based primarily on completeness of exposure and geographic location, and secondarily on structural simplicity and accessibility. Numerous sections in nearby mountain ranges were also studied to obtain an understanding of regional stratigraphic relationships.

PREVIOUS WORK

Until 1960, published stratigraphic and paleontological studies of Late Paleozoic sediments from southern Nevada were few. Since 1961 interest has grown substantially as reflected by an increase in the number of papers dealing with the area.

The earliest report of Carboniferous strata in southern Nevada was published by Spurr in 1903. Longwell (1921, 1928) described the geology of the Muddy Moun-

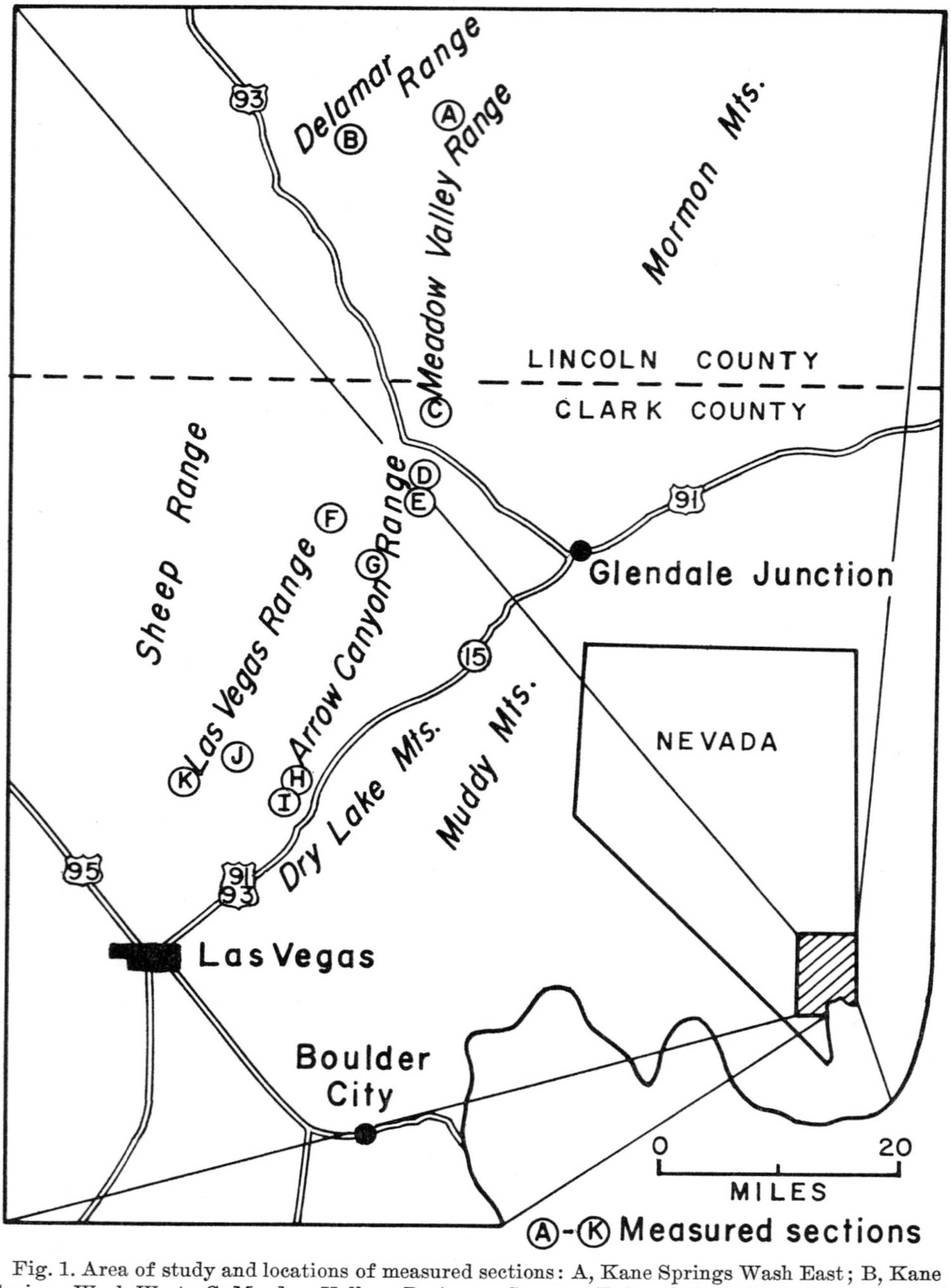

Fig. 1. Area of study and locations of measured sections: A, Kane Springs Wash East; B, Kane Springs Wash West; C, Meadow Valley; D, Arrow Canyon; E, Battleship Wash; F, Las Vegas Range; G, Dry Lake; H, Arrowlime Quarry; I, Apex; J, Gunnery Range; and K, Gass Peak.

tains, proposing names for formations in the earlier work. Hewett (1931) mapped the Goodsprings area and assigned rock unit names there. From studies in the Spring Mountains, Longwell and Dunbar (1936) subdivided the Bird Spring formation into members. Geographic distribution of strata in Clark County was mapped by Bowyer, Pampeyan, and Longwell (1958) and modified by Longwell and others (1965). A geologic map of Lincoln County by Tschanz and Pampeyan was published in 1961. Rich (1961, 1963) identified and illustrated the fusulinids and the carbonate petrography of the Bird Spring formation at Lee Canyon. Langenheim and others (1962) described the section in the Arrow Canyon Range. Brill (1963) correlated Permian and Pennsylvanian stratigraphy of the western Colorado Plateau with that of the eastern Great Basin. Coogan (1964) discussed the Early Pennsylvanian history of the Ely Basin and correlated these rocks with those of the Bird Spring Basin of southern Nevada. Langenheim and Langenheim (1965) described the Chester through Wolfcamp section at Arrow Canyon. Heath, Lumsden, and Carozzi (1967) did a petrographic study of the Morrow into lower Missouri section at Arrow Canyon.

Descriptions of some Late Paleozoic fossils from southern Nevada are included in papers by Lane (1961, 1963, 1964*a*, 1964*b*), McCutcheon and Wilson (1961), Wilson, Waines, and Coogan (1963), Macurda (1964), Dunn (1965, 1966), Cassity and Langenheim (1966), Lane and Webster (1966), and Webster and Lane (1967).

ACKNOWLEDGMENTS

I wish to thank the following: Dr. N. Gary Lane for initiating my interest in the area and for uncounted suggestions, stimulating discussions, and helpful criticisms; Standard Oil Company whose Fellowship in Geology for 1962–63 provided me with field expenses; Dr. Carl B. Rexroad for reviewing conodont identifications and offering helpful comments; Mr. and Mrs. Chris Pedersen and many other permanent and temporary residents of the Muddy River valley for generous hospitalities extended; and Mrs. Opal Kurtz and Miss Julie Guenther for drafting the illustrations.

STRATIGRAPHY

INTRODUCTION

Formations investigated include part or all of the Chainman shale and the Battleship Wash, Indian Springs, and Bird Spring formations. The Chainman, a black shale of Mississippian age, occurs in the northern part of the area. The Battleship Wash is a gray, bioclastic limestone occurring in the central and southern part of the area. It is equivalent at least to the upper part of the Chainman shale. Total thicknesses of these two formations were not measured. Both are overlain by the Indian Springs formation. The Indian Springs formation (Mississippian age) is an interbedded, red-weathering sequence of shale, limestone, orthoquartzite, and conglomerate; it averages 180 feet in thickness. The Bird Spring formation (Late Mississippian to Permian in age) conformably overlies the Indian Springs and is dominantly gray bioclastic limestone. Only the Morrow and Derry parts of this formation were investigated. The stratigraphy of the eleven sections measured is illustrated in plate 1 (in pocket) and is diagrammatically summarized in figure 2.

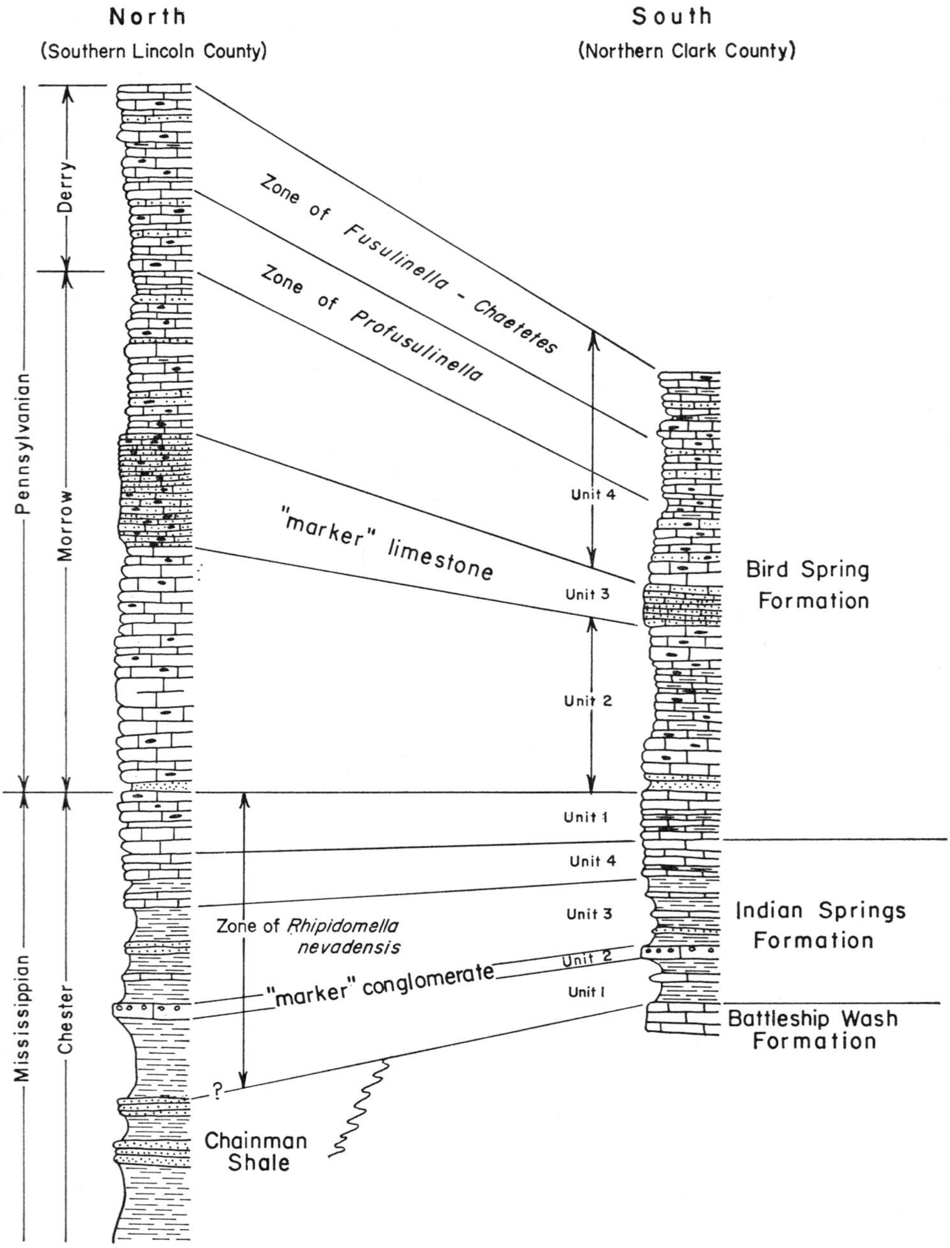

Fig. 2. Stratigraphic relationships of Upper Mississippian–Lower and Middle Pennsylvanian formations to informal rock units in the study area. Not to scale.

Chainman Shale

The Chainman shale, named by Spencer (1917), was described by Sadlick (1960) as a clastic wedge derived from the west, thickest in central Nevada, and thinning to the east. It is characterized by gray to black platy shale that weathers greenish-black to gray and may contain clean, ferruginous-brown weathering, fine-grained orthoquartzite beds in its upper part.

Langenheim (1963*b*) named the Hamilton Canyon formation for the sandy shale that had previously been considered as the upper part of the Chainman. He recognizes (1963*b*, fig. 2) a thickness of 270 feet for this formation at Kane Springs Wash, above the Chainman and below Pennsylvanian limestones. Webster and Lane (1967, fig. 2) show a thickness of 240 feet of Indian Springs formation at Kane Springs Wash, above the Chainman and below the Bird Spring formation. Furthermore Webster and Lane show sandstone beds in the upper part of the Chainman as they recognize it. Obviously the Hamilton Canyon as recognized by Langenheim is equivalent to the uppermost part of the Chainman and the Indian Springs as recognized by Webster and Lane. Webster and Lane demonstrated that the Indian Springs is widespread in southern Nevada and may be traced lithologically and faunally from the type section at Indian Springs to Kane Springs Wash. One lithologic unit, the "marker" conglomerate of Webster and Lane (p. 510) was reported from all sections north of the Las Vegas Valley but not recognized at Indian Springs. Although poorly exposed, this unit has since been found by them at Indian Springs and Lee Canyon. Because the Indian Springs is clearly identifiable and has priority, the Hamilton Canyon is not recognized at Kane Springs Wash.

At Kane Springs Wash (Section A) the Chainman yielded an abundant goniatite fauna with numerous specimens of *Cravenoceras hesperium* Miller and Furnish and rare specimens of *Eumorphoceras* sp. from beds near the middle of the formation, and a few specimens of *Cravenoceras merriami* Youngquist from shale in the upper part. Those goniatites indicate a Chester age for the formation. Although a complete section of Chainman shale is exposed near Section A (Kane Springs Wash East), only the upper 162 feet were measured.

In the northern part of the study area, the Chainman grades upward into the conformably overlying Indian Springs formation (Webster and Lane, 1967). The top of the former is placed arbitrarily at the top of the highest tan-weathering orthoquartzite bed below the "marker" conglomerate (unit 2) in the Indian Springs. In northern Clark County the Battleship Wash formation is correlative, at least in part, with the Chainman; the change in facies between these two formations is exposed approximately 10 miles south of Section A in the Meadow Valley Mountains.

Orthoquartzite beds in the upper part of the Chainman are probably correlative with the lower part of the Scotty Wash quartzite near Pioche, Nevada. In the Pioche area, Westgate and Knopf (1932) described the Scotty Wash as a reddish-brown, quartzitic sandstone. Because quartzite beds are subordinate to the dominantly shale lithology at Kane Springs Wash they are included in the Chainman shale and the Scotty Wash quartzite is not recognized.

Battleship Wash Formation

The Battleship Wash formation was called "BS_a" by Langenheim and others (1962) and subsequently named by Langenheim and Langenheim (1965) from exposures in the Arrow Canyon Range. It consists of resistant bluish-gray weathering, medium- to thick-bedded, bioclastic limestone with a few thin arenaceous beds in the lower part. These arenaceaus beds may be correlative with thin orthoquartzites in the upper part of the Chainman shale farther north at Section A (Kane Springs Wash East).

Langenheim and Langenheim (1965) include the Battleship Wash in the Bird Spring Group and consider (ibid., p. 299) that their "BS_a" and "BS_b" units together are equivalent to the Indian Springs member of the Bird Spring formation. From the original description and graphic column of Longwell and Dunbar (1936) it is clear that they did not recognize the Battleship Wash lithology in the Spring Mountains and therefore did not include it in their Indian Springs member. Field study of Indian Springs exposures in the Spring Mountains and at Indian Springs, Nevada, revealed that the Battleship Wash Formation is present south of the Las Vegas Valley, but very thin and only at Lee Canyon. At all other localities the Indian Springs paraconformably overlies the Monte Cristo formation, the Battleship Wash formation being absent. Throughout the Las Vegas and Arrow Canyon ranges the lowest Indian Springs lithology is either a thin orthoquartzite or varicolored-to-black shale that appears to conformably overlie the Battleship Wash formation. Because of the abrupt lithologic change and an obvious faunal break at the top of the Battleship Wash, a slight hiatus is believed to have occurred in the area prior to deposition of the Indian Springs member. Therefore, the Battleship Wash is judged to be neither a part of the Bird Spring formation nor equivalent to part of the Indian Springs, but rather to be an easily recognizable, lithologic entity.

Some doubt exists about the usefulness of the name Battleship Wash. Longwell (1921) applied the name Bluepoint limestone to a poorly defined section of rocks in the Muddy Mountains. Girty, in Longwell's report, assigned a Chester age to the formation based on fossils collected by Longwell. Langenheim and Langenheim (1965, p. 227) report that Longwell's fossil collections are lost or mislaid. Furthermore, they state (p. 229): "The BS_a unit is of the proper lithology, age, and stratigraphic position for assignment to the Blue Point Limestone, but the latter was not adequately defined and definite correlation appears to be impossible." If correlation of these two units is proven, then the Bluepoint has priority. Until this matter is resolved, it is recommended that the Battleship Wash formation be recognized in the Arrow Canyon, Las Vegas, and Meadow Valley ranges and the Bluepoint limestone in the Muddy Mountains.

Although not measured for this study, the Battleship Wash formation was used as the base of sections C (Meadow Valley) through K (Gass Peak) and the upper 1 foot was collected for fossils. Except at sections H (Arrowlime) and I (Apex) in the southern Arrow Canyon Range, a megafauna that included *Flexaria, Michelinia, Spirifer,* solitary rugose corals, and fenestrate bryozoans was commonly found on the uppermost bedding surface (figs. 4–8, in pocket). In addition

Rhipidomella nevadensis (Meek) and fragments of *Lepidodendron* were found on this surface at Section D (Arrow Canyon). With only minor variations the same abundant and diverse conodont fauna was found in this bed at all sections. An early Chester age is suggested for this formation by the conodonts. The absence of megafossils in sections H (Arrowlime) and I (Apex) is explained by a lack of surface exposure at Section H and dolomitization at Section I.

Indian Springs Formation

Longwell and Dunbar (1936) named the Indian Springs, designating it as the lowest member of the Bird Spring formation. They included some 700 feet of strata lying unconformably on the Lower and Middle Mississippian Monte Cristo limestone and beneath the *Fusulinella* Zone. Webster and Lane (1967) elevated the Indian Springs to formational rank because it is lithologically distinct, weathering rust-brown to red-orange, topographically expressed as a strike valley, and laterally persistent, forming a good, mappable unit (pl. 2). They restricted it to the basal 170 feet of strata at Longwell and Dunbar's type section at Indian Springs, Nevada. The Indian Springs formation is used as emended in this report.

The Indian Springs is present throughout the area studied and it ranges in thickness from 132 feet at Section J (Gunnery Range) to 250 feet at Section F (Las Vegas Range), generally tending to thin to the south and east. Lithologically it may be subdivided into four units.

Unit 1, the lowest, is 90 percent black shale with one to five discontinuous, thin, ferruginous bioclastic limestones and one to three tan- to purple-weathering orthoquartzites. It ranges from 31 to 123 feet thick. A complete exposure, 95 feet thick, of unit 1 occurs in the south wall of Arrow Canyon where Welsh (1959) found one specimen of *Cravenoceras* sp. No fossils were found by me in either the shale or the quartzites; the quartzites show oscillation ripple marks and small-scale cross-bedding. In addition to a diverse but not abundant conodont fauna the limestones contain *Rhipidomella nevadensis, Schizophoria, Flexaria, Anthracospirifer,* an orbiculoid brachiopod, and unidentified ramose bryozoans and solitary corals.

Unit 2 conformably overlies unit 1 and is a resistant tan- to gray-weathering bioclastic limestone 2 to 6 feet thick with a 0.5- to 1.5-feet thick, dark red- to black-weathering calcareous- and ferruginous-cemented conglomerate in the upper part. The fauna of this unit is the same as in unit 1, and in addition contains poorly preserved, unidentifiable internal molds of pelecypods, gastropods, and rare cephalopods in the conglomerate. Conodonts are abundant in the matrix of the conglomerate. The clasts in the conglomerate are pebbles and cobbles of quartzite and chert, with rare pieces of the underlying limestone. Because this unit is lithologically distinct, laterally persistent, and contains a distinct fauna it is referred to as the "marker" conglomerate by Webster and Lane (1967).

Unit 3 is 47 to 107 feet thick and composed of shale interbedded with limestone and orthoquartzite that make up 30 to 50 percent of the unit. Although recognizable as a unit, individual beds within the unit are discontinuous and cannot be correlated from section to section. Shales are dominantly black in color, although gray to green and occasionally varicolored shales are present. Few fossils were found within the shales except for bed 16, Section G (Dry Lake), which yielded an extremely abundant megafauna of the *Rhipidomella nevadensis* Assemblage Zone, including species of *Archimedes, Pentremites, Inflatia, Flexaria,* and *Schizophoria,* among others. Limestones are thin- to medium-bedded, coarsely bioclastic, and weather red-brown. The lowest limestone of unit 3 at Section I (Apex, bed 8) contains a cephalopod fauna including *Syngastrioceras walkeri* Webster and Lane, *Pronorites* sp., *Cravenoceras* sp., *Liroceras liratum* (Girty) and *Mooreoceras* sp. This fauna suggests a late E_2 age (upper *Eumorphoceras* Zone of Chester age) for the bed. Other limestones in unit 3 contain elements of the *R. nevadensis* assemblage. Orthoquartzites are fine-grained, with small-scale cross-bedding and commonly contain fragments of *Calamites.* Limestones become increasingly abundant upwards and unit 3 grades into unit 4.

Limestones and subordinate orthoquartzites and thin, light- to dark-gray shales comprise unit 4, which ranges from 40 to 66 feet thick. The limestones and shales yield abundant fossils of the *Rhipidomella nevadensis* assemblage. Channeling was observed at the base of this unit in Section D (Arrow Canyon, bed 15) where a pebble-sized coquina containing rounded and worn fossil fragments has cut into the underlying sandstone of unit 3.

Units 1 to 4 are present throughout the area except at Section B (Kane Springs Wash West), where units 1 and 2 are not exposed because of faulting. All units show the same general trend of thinning to the south and east as was noted for the formation. These units are recognizable at Lee Canyon in the Spring Mountains and at the Indian Springs type section. Because the four units described above are all of middle to late Chester age, typically poorly exposed (especially units 1 and 3), and too thin to map at scales normally used in the Great Basin, no formal nomenclature is applied to them.

Bird Spring Formation

The Bird Spring formation (Hewett, 1931) is lithologically similar to the Callville limestone (Longwell, 1921) of the Muddy Mountains and clearly lacks the priority of the latter. As pointed out by several authors (Longwell and Dunbar, 1936; Welsh, 1959; Rich, 1961; Brill, 1963; Coogan, 1964; Longwell and others, 1965; Langenheim and Langenheim, 1965), however, the Bird Spring thickens northwest from the type section and it is convenient to apply the term to the basinal or thicker facies and to restrict the Callville to the shelf facies, using the Wasatch–Las Vegas Line as the nomenclatorial and facies boundary.

Longwell and Dunbar (1936) divided the Bird Spring into five members based on lithology and fusulinid content. Except for the Indian Springs member discussed above, none of the members were recognizable on a lithologic basis in the Las Vegas, Arrow Canyon, or Meadow Valley ranges, although time-equivalent strata are recognized by fusulinids. Elevation of the Indian Springs to formation rank and its exclusion from the Bird Spring restricts the latter to gray-weathering limestones previously included in the Indian Springs but below the *Fusulinella* Zone and to the upper four numbered members of Longwell and Dunbar.

Langenheim and others (1962) divided the Bird Spring formation at Arrow Canyon into five units, BS_a through BS_e, and elevated it to a group. These same divisions were recognized by Langenheim and Langenheim (1965) when they named the BS_a the Battleship Wash formation, which should not be included in the Bird Spring as pointed out above. Their BS_b is equivalent to the Indian Springs formation as restricted by Webster and Lane (1967), and the BS_c through BS_e are recognizable only with difficulty in Arrow Canyon and could not be recognized in nearby sections. Therefore, these units are considered to be informal and elevation of the Bird Spring formation to group status is not accepted.

This study of the Chester through Derry part of the Bird Spring formation disclosed only two distinct lithologies that could be traced throughout the area of investigation: a clastic sequence in the basal Pennsylvanian strata and a sandy limestone in the middle Morrow sequence. Neither of these distinctive lithologies could be found in surrounding areas. Utilizing these two key beds, the Chester through Derry part of the Bird Spring was divided into four informal units.

Unit 1 is 20 to 62 feet thick, consists of thin- to medium-bedded, gray, bioclastic limestones with dark-brown to black nodules and discontinuous bands of chert and thin beds or partings of light-gray shale. It contains abundant fossils of the *Rhipidomella nevadensis* Assemblage Zone comparable with those of the upper part of the Indian Springs formation. Shale partings especially yield *R. nevadensis* in abundance. The assemblage implies a late Chester age for these beds.

Unit 2 ranges from 183 feet at Section I (Apex) to 543 feet thick at Section A (Kane Spring Wash East), thinning to the southeast. It is best exposed in Arrow Canyon (Section D) where it is 362 feet thick (pl. 3, fig. 1). This unit contains a basal clastic sequence in all sections and is overlain by a thick sequence of carbonates of two facies, a northern and a southern, which are interbedded in the central part of the Arrow Canyon Range.

The basal clastic sequence ranges from 2 to 21 feet thick. At sections D (Arrow Canyon, beds 29–31) and I (Apex, beds 19–21) the lowest clastic bed is a pebble conglomerate overlain by a yellow shale in Section D and by a yellow quartzite in Section I followed by limy sandstone and sandy shale, respectively. At sections C (Meadow Valley, bed 13), H (Arrowlime, bed 22), and K (Gass Peak, bed 22), the clastic sequence is sandstone that is overlain by limestone. All other sections have a covered interval, suggestive of shale, followed by sandy limestone. These clastic beds are thought to represent deposition after a brief hiatus during latest Mississippian or earliest Pennsylvanian time. Fossils were not found in these beds. Because the clastics grade into the overlying limestones, they are included in the Pennsylvanian part of the Bird Spring. Two facies are recognized in limestones overlying the clastic beds: a northern one that includes sections A through E (from Kane Springs to Battleship Wash) and a southern carbonate facies covering sections F through K (from Las Vegas Range to Gass Peak).

The northern facies consists of cliff-forming, dark-gray to black, medium- to thick-bedded, cherty limestones in the lower part grading upward into bench-slope, weathering, interbedded, cherty limestone and argillaceous limestone with a few thin calcareous shales. Rare sandy limestones are found throughout the facies. Megafossils, dominantly brachiopods, are found occasionally throughout this facies and are especially abundant in a shale in Arrow Canyon (Section D, bed 46). Species of *Composita, Orthotetes, Anthracospirifer,* and productoids are most common. They appear to range through the interval and are of little stratigraphic value. *Schizophoria resupinoides* (Cox) and *S. texana* (Girty) are moderately abundant in the lower cliffy beds and a few specimens of *Rhipidomella eleyensis* Lane occur along shaly partings in the upper bench-slope beds. Conodont species that are found in the lower cliffy beds range through unit 2 except for two species, *Gnathodus defectus,* which was not found above the basal few feet of the unit, and *G. muricatus,* which was not found above the lower one-third of the unit. Four species, *Gnathodus bassleri, Streptognathodus suberectus, S. expansus,* and *Idiognathodus delicatus,* were not found below the bench-slope part of unit 2. *Streptognathodus suberectus* is apparently restricted to the lower two-thirds of the bench-slope weathering beds; *S. expansus* ranges through the bench-slope beds and into the lowest part of unit 3; and *G. bassleri* and *I. delicatus* range through the Derry.

The southern facies of unit 2 (sections F, Las Vegas Range, through K, Gass Peak) consists of a lower 20- to 30-feet thick, cliffy, sequence similar in lithology and faunal content to the lower half of the northern facies. This cliff-forming sequence is overlain by black, platy, thin-bedded, argillaceous, cherty limestone that grades upward into less argillaceous bioclastic limestones. The argillaceous limestones grade into black, fissile shale at Section J (Gunnery Range, bed 17) and into red, platy shale at Section I (Apex, bed 25). The fauna from the argillaceous beds is abundant and dominated by brachiopods, of which *Lissochonetes* is most abundant. Conodonts are not affected by the carbonate facies change and show the same stratigraphic distribution in the southern as in the northern facies.

Above the limestones that exhibit facies change, the upper part of unit 2 contains one to four biostromal beds of dendroid corals in all sections except F and G (Las Vegas Range and Dry Lake). These beds are formed by a maze of coralla questionably identified as *Tschussovskenia,* a genus previously known only from Upper Carboniferous and Permian strata of Russia. In Section A (Kane Springs Wash East, bed 33) four coral biostromes 2 to 3 feet thick occur 116, 108, 103, and 90 feet below the top of unit 2. At Section B (Kane Springs Wash West, bed 40) the upper three of the four biostromes are present at 14, 11, and 5 feet below the top of unit 2. Section C (Meadow Valley, bed 22) contains two biostromal beds 24 feet apart which could not be definitely correlated with individual beds of sections A and B, but the upper one is probably correlative with the top biostromes of sections A and B because all contain associated specimens of *Multithecapora* not found in the lower beds. A single biostrome was found in all other sections. At each of these occurrences numerous colonies of *Multithecapora* were found in association with *Tchussovskenia?;* therefore, they are correlated with the top biostrome at sections A, B, and C. Because of fossil content and stratigraphic position, an early and middle Morrow age is assigned to unit 2.

Unit 3 ranges from a featheredge (sections I, Apex, and K, Gass Peak) to 108 feet thick (Section A, Kane Springs Wash East, bed 36). The unit exhibits only minor variations in thickness north of Section G (Dry Lake), but thins abruptly both southeast and southwest of that section. Lithologically this unit consists of light-gray, thin-bedded, very arenaceous, cherty, finely crystalline limestone that weathers as a light-brown massive unit. The thin-bedded character is noticeable only upon close inspection. The unit is usually well exposed and forms a formidable cliff where dips are less than 20°. Thin shales are present approximately 25 feet below the top of the unit at sections A (Kane Springs Wash East) and D (Arrow Canyon). Few megafossils were found in unit 3 and the conodont species are the same as in the upper part of unit 2. A middle Morrow age is assigned to the unit. Because of the characteristic lithology and prominent exposure this unit is referred to here as the "marker" limestone.

Unit 4 ranges from 482 feet (Section I, Apex) to 828 feet thick (Section G, Dry Lake), and exhibits greater irregularities in thickness than units below. It is thinnest to the southeast and from Section G thins toward Section D (Arrow Canyon, 591 feet), then thickens farther to the north to 682 feet at Section B (Kane Springs Wash West). Section A (Kane Springs Wash East) lacks the upper part of unit 4 because of faulting. Lithologically unit 4 consists of interbedded thin- to medium-bedded, gray-weathering, bioclastic limestone; similar limestone containing nodules and discontinuous bands of chert; argillaceous and arenaceous limestone. The unit weathers in a bench-slope relief (pl. 3, fig. 2) and no single lithology was found which could be traced throughout the eleven measured sections.

Using distinctive faunal characteristics unit 4 may be subdivided into three parts. First, a lower upper Morrow interval extends from the top of the "marker" limestone to the base of the *Profusulinella* Zone. This sequence contains a distinctive, thick-plated crinoid, *Paragassizocrinus calyculoides* (Lane) n. comb., which is found in sections D (Arrow Canyon) and F (Las Vegas Range) through J (Gunnery Range) from just above to 160 feet above the "marker" limestone. Other megafossils are not abundant and are dominantly long-ranging species of brachiopods, similar to fossils above and below. No conodont species were found to be restricted to this sequence. On the basis of stratigraphic position and the few diagnostic fossils found, a late Morrow age is assigned to this interval.

Next above is the zone of *Profusulinella* of early Derry age. Specimens of *Profusulinella* were found only at sections C, D, E, and G (the central part of the study area). Conodont faunas found in sections B, F, I, and K are equivalent to ones found within the *Profusulinella* Zone at sections that yielded fusulines. This indicates that the *Profusulinella* Zone is present in all sections investigated. One species of conodont, *Neoprioniodus? expandofundus*, n. sp., was found only within the *Profusulinella* Zone. No specimens of *Chaetetes* were found in this zone as reported by Coogan (1964).

Third, the faunizone of *Fusulinella-Chaetetes* ranges from 146 feet (Section D, Arrow Canyon) to 202 feet thick (Section K, Gass Peak). In addition to abundant specimens of *Fusulinella* this zone includes three to five beds containing few to numerous heads of the tabulate coral *Chaetetes* (fig. 3). At sections C (Meadow Valley) through G (Dry Lake) the first occurrence of *Fusulinella* was found on the upper surface of the lowest bed containing *Chaetetes*. In sections B (Kane Springs Wash West) and H (Arrowlime) through K (Gass Peak) the lowest *Fusulinella* was found 25 to 100 feet above the lowest *Chaetetes*. At sections D (Arrow Canyon) and E (Battleship Wash) the highest occurrence of *Fusulinella* is 30 feet below the highest *Chaetetes*, and *Fusulina* is found 15 feet above the latter. At these localities the highest *Chaetetes* is associated with a large phaceloid, rugose coral identified as *Caninostrotion?*. In sections C (Meadow Valley), G (Dry Lake), J (Gunnery

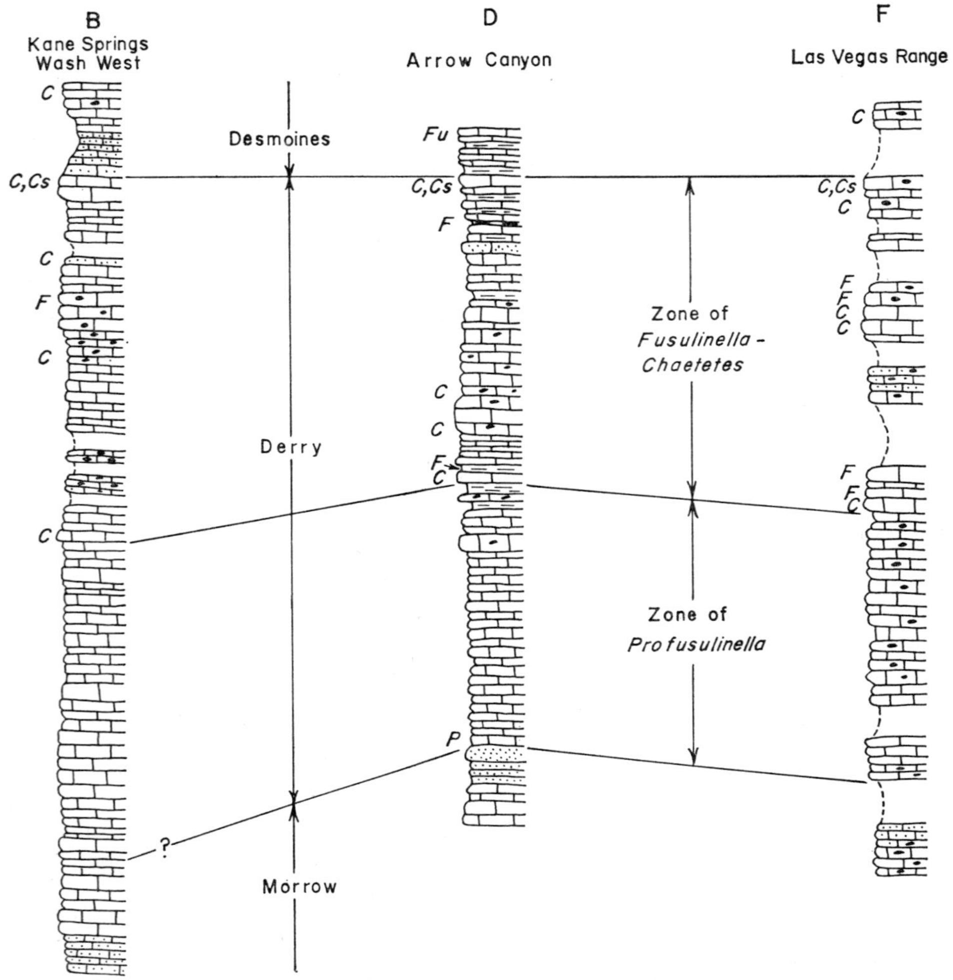

Fig. 3. Distribution of *Chaetetes* (*C*), *Caninostrotion* (*Cs*), and fusulines (*Profusulinella* [*P*]; *Fusulinella* [*F*]; *Fusulina* [*Fu*]) in Middle Pennsylvanian rocks of southern Nevada. Not to scale.

Range), and K (Gass Peak) the same association is found in the highest *Chaetetes* bed. Sections B (Kane Springs Wash West), F (Las Vegas Range), and H (Arrowlime) have one higher, *Chaetetes*-bearing bed 45, 45, and 22 feet, respectively, above the *Chaetetes-Caninostrotion?* layer. Only rare *Chaetetes* coralla are found in these higher beds. Because *Fusulinella* is associated with the lowest *Chaetetes* bed at several localities, *Chaetetes* and *Fusulinella* are interbedded, and the *Chaetetes-Caninostrotion?* horizon is between the highest *Fusulinella* and below the lowest *Fusulina,* the interval common to *Fusulinella* and *Chaetetes,* is called the *Fusulinella-Chaetetes* Faunizone.

Dott (1955) referred to the *Chaetetes-Profusulinella* Faunizone in northeastern Nevada; Coogan (1964) noted the occurrence of *Chaetetes* and *Profusulinella* in the Ely area; Brill (1963) reported *Chaetetes* in Desmoines rocks in the Spring Mountains; and I have observed abundant *Chaetetes* within the *Fusulina* Zone in the Paradox formation along the San Juan River. Thus, the range of *Chaetetes* in the Great Basin is Derry and Desmoines. The *Fusulinella-Chaetetes* faunizone is recognized only in southern Nevada and because of the presence of *Fusulinella*, is assigned a late Derry age. No conodont species were found that are unquestionably restricted to this zone.

FAUNA

Introduction

As noted in the formational descriptions, certain fossils were found to be useful in subdividing units, determining stratigraphic position, and correlating sections. Megafossil identifications are tabulated for collections from each section in figures 4 through 8 (in pocket); conodont identifications are tabulated in figures 9 (in pocket) through 15. These figures show the relative representation of each phylum by number of recognized genera. Five phyla: Coelenterata, Bryozoa, Brachiopoda, Mollusca, and Echinodermata, and the order Conodontophorida have the most numerous genera and each of these would make a worthwhile systematic investigation. All collected fossils are, however, too diverse for this paper. Therefore, detailed paleontologic studies here are restricted to conodonts. This group was chosen because of its relative abundance and importance in correlation with Carboniferous strata in Europe and the midcontinent. A few generalities concerning other groups not treated taxonomically below are discussed here.

Foraminifera are common in shales throughout unit 4 of the Indian Springs formation and the entire Bird Spring formation. The *Profusulinella* and *Fusulinella* zones are of special significance and are correlated with Derry strata of the midcontinent.

Sponge spicules are found in several beds of the Bird Spring formation but appear to be of no stratigraphic value.

Coelenterates are common and some beds contain an abundance of one or two taxa. Species of *Lophophyllidium, Michelinia*, and *Syringopora* are found throughout the sequence studied and are of no apparent stratigraphic value at the generic level. The *Tschussovskenia*? biostromal beds may be an index for strata of middle Morrow age in southern Nevada. Occurrence of *Chaetetes* interbedded with *Fusulinella* appears to delimit a recognizable faunizone in southern Nevada. The association of *Caninostrotion*? with *Chaetetes* may be another good marker in southern Nevada and this association should be looked for in surrounding areas.

Bryozoans are numerous and usually well preserved in shales and limestones of the Indian Springs and Bird Spring formations. *Archimedes* is present in unit 3 and abundant in unit 4 of the Indian Springs and unit 1 of the Bird Spring. This genus is a Mississippian guide fossil in the midcontinent. Large, well-preserved specimens of an unidentified massive fistuliporid bryozoan are common in units 3 and 4 of the Indian Springs. A large unidentified ramose bryozoan with a zoarium approximately 1 inch in diameter was found commonly in the upper Morrow and lower Derry part of the section. Detailed study of this group is needed in Carboniferous strata of southern Nevada.

Nearly fifty species of brachiopods, referred to 33 genera, were found in the 11 sections studied. Throughout these sections productoids and spiriferids are the most common and usually are well preserved. A systematic investigation of the brachiopods will probably allow designation of a series of assemblage zones for Lower and Middle Pennsylvanian strata in southern Nevada. Sadlick (1954) drew attention to the importance of the *Rhipidomella nevadensis* Faunizone. He has shown the widespread occurrence of this zone and its stratigraphic usefulness in other works (Sadlick, 1955, 1957, 1960, 1963). In southern Nevada associated brachiopods in this assemblage zone include species of *Inflatia, Flexaria, Schizophoria,* and *Anthracospirifer,* among others. *Inflatia* and *Flexaria* are not known in strata younger than Mississippian.

Mollusks are not common throughout the sections but are abundant at specific localities. Gastropod and pelecypod genera are long-ranging forms typical of Late Paleozoic strata. Scaphopods were found in moderate abundance at sections G (Dry Lake) and H (Arrowlime Quarry) in the basal part of unit 4 of the Bird Spring Formation. Webster and Lane (1967) showed that the cephalopods found in the Indian Springs Formation suggested correlation with the E_2 Zone of Europe.

Echinoderm remains, although seldom found articulated, are common in Late Paleozoic rocks in southern Nevada. Two classes, Blastoidea and Crinoidea, were studied in detail; these studies are being prepared for publication. The blastoids are found only in the late Chester part of the Indian Springs and Bird Spring formations. They show affinities with described species of Late Mississippian age of the midcontinent. All articulated crinoids are species of described inadunates and suggest a Late Mississippian to Middle Pennsylvanian age. Echinoid spines and disarticulated plates are common in the Derry part of the Bird Spring formation. Unidentified fragments of trilobites are found sporadically throughout the area. Fish teeth are especially abundant in the upper part of the Indian Springs and basal part of the Bird Spring formations but are found throughout the section.

Two groups of plants are found in the sections studied and are of stratigraphic importance. Silicified logs of *Lepidodendron* are found on the uppermost bedding surface of the Battleship Wash formation at Arrow Canyon. Fragments of *Calamites* and occasionally *Stigmaria* are present in the sandstones and orthoquartzites in the Indian Springs formation. Algae including *Osagia* and *Komia* are present in the Morrow and Derry part of the Bird Spring.

CONODONTS

Ranges and relative abundances of Chester through Derry conodonts found during this study are shown in figure 9 (in pocket). Thirty-eight species (three of which are new) belonging to twelve genera are identified and an additional six genera are recognized from fragments. Platform types are the most abundant and useful for correlation. Blade types, the second most common form, are of stratigraphic value, especially in the Chester part of the section. Bar types are usually fragmentary and of limited stratigraphic value. No conodonts were found in the Chainman shale. Species distribution and relative abundance for each of the six sections sampled for conodonts are shown in figures 10 through 15.

Bed Number \ species	*Cavusgnathus naviculus*	*C. convexus*	*C. spathus*	*C. gigantus*	*C. lautus*	*Gnathodus bilineatus*	*G. girtyi simplex*	*G. defectus*	*G. muricatus*	*G. bassleri*	*Streptognathodus unicornis*	*S. expansus*	*S. noduliferus*	*Idiognathoides convexa*	*Idiognathodus delicatus*	*Spathognathodus* sp.	*S. minutus*	*Neoprioniodus? expandofundus*	*N. loxus*	*N.* cf. *camurus*	*Ozarkodina delicatula*	*Synprioniodina microdenta*	*Metalonchodina* sp.	*Ligonodina* spp.	*Lonchodina* spp.	*Hindeodella* spp.
76										R					R											
75										F					R											
74																										F
73				R																						
72				R																						
71										A					A											
70															R											
69										R																
68										F					C											R
65				F						F					A											
63				R						A					C			R						R	R	
62															F											
61										C					C											
60				F						A					C											
59										F					A						R					
58					F										R											
56				C	C					F					A											R
54				F											R											
50				F																						
47													F		C				R							
45				R	F														R							
41													R													
40			C									R	A	F	A											C
39					F																					R
37			F	R								C	A	C												R
35			R	R	F					F			A	F											R	
34				R									F													
33					R								A	F												
32			C	F									R													C
29									R																	F
28			F	F					C																	R
27				F					A																	R
26				F	F				A																	
24			F					F	C		F															F
23					R				C		R															
22				C																						
21							F				R															
20						F	F				F						R				R			R		F
18							R																			
16		R																								
14						A	C																R		C	C
12																										R
9							C																			
8							F									F				R	R	R				F
6		F				R					F															R
4		A									R															
2																	R									
1	R										F															

Fig. 10. Species distribution and relative abundance for Section B, Kane Springs Wash West; locality 5243. Abbreviations: R=1, F=2–4, C=5–10, A=>10 specimens respectively.

Bed Number \ species	*Magnilaterella robusta*	*Cavusgnathus altus*	*C. convexus*	*C. spathus*	*C. gigantus*	*C. lautus*	*Gnathodus commutatus*	*G. bilineatus*	*G. girtyi simplex*	*G. muricatus*	*G. defectus*	*G. opimus*	*G. bassleri*	*Streptognathodus unicornus*	*S. lanceolatus*	*S. suberectus*	*S. expansus*	*S. noduliferus*	*Idiognathoides convexa*	*Idiognathodus claviformis*	*I. delicatus*	*Spathognathodus rexroadi*	*S. minutus*	*Neoprioniodus singularis*	*N. scitulus*	*N. loxus*	*Ozarkodina curvata*	*O. roundyi*	*O. delicatula*	*O. campbelli*	*Synprioniodina denticamura*	*Hibbardella ortha*	*Roundya* sp.	*Metalonchodina* sp.	*Ligonodina* spp.	*Lonchodina* spp.	*Hindeodella* spp.
108					R								F								F		R														
107													R								F																
105													F								F																R
104						F																															
103													F								F								R								C
102					C	C							C								A		F														C
101													R																								
99												F	F								A									R							R
98					C	F															C																R
96																					F																
93					C	F							A								C		R														
91												R	R								C																
88						R							A							A	A		F							R							C
87						F							C								A																
84						F						C									C																F
82					F	F															F																
81																					A																
80					F	F						R	F								A		F						F								
75					C	C															A																
73													F								C																
72													R								A																F
71					A	A															F									C							C
66						F							R						R		A												R		A	C	A
65																																					R
63				R	C	F																														R	R
62					R																C																
61						F															F															R	R
59					F	F												R			F																F
58																					F																F
57				F	F	R										A	A	A	R		F																F
56																	F	A	F		R																C
53																		F			R																
51																		A																			F
50																		A	C																		
49				R	R											C	C	A	F																		
47					F	R									F			A											R							R	R
45				F		C												A	R																		R
43				F	F	C												A								R											F
42																																					F
41					F													C	F																		
40					R	R					C							C	C																		F
36					F	R					A								R													R					R
35					R	F																															R
33				F	F	C																				R			R						F		F
32				R	A	A				C	A							F	F																A	A	
28					R					A	C																		R								R
27											R																										
26a										F	C			R																							
26					F	F			R		F			F																							
24							R	A	A	F	F																		F							R	R
23																																					R
20			R								R																										F
15								R																													
13			R								R			F																							
9								A	F																									R	F	A	C
7							F	C						F									R														C
4		C	F											F																							F
1	R	A					C		A													A		F	F	R	F	R	C		R				R	F	A

Fig. 11. Species distribution and relative abundance for Section D, Arrow Canyon; locality 5245. Abbreviations: R=1, F=2–4, C=5–10, A=>10 specimens respectively.

Bed Number \ species	*Geniculatus?* sp.	*Cavusgnathus altus*	*C. naviculus*	*C. spathus*	*C. gigantus*	*C. lautus*	*Gnathodus commutatus*	*G. bilineatus*	*G. girtyi simplex*	*G. defectus*	*G. muricatus*	*G. opimus*	*G. bassleri*	*Streptognathodus unicornus*	*S. noduliferus*	*Idiognathoides convexa*	*Idiognathodus delicatus*	*Spathognathodus rexroadi*	*S. minitus*	*S. coloradoensis*	*Neoprioniodus singularis*	*N. scitulus*	*N. loxus*	*Ozarkodina roundyi*	*O. curvata*	*O. delicatula*	*O. campbelli*	*Hibbardella ortha*	*Roundya* sp.	*Ligonodina* spp.	*Lonchodina* spp.	*Hindeodella* spp.
90						F							C				A									R						
87													F				F															
81					R	F						F	C				A		R	R										R		F
79					C	F											A															C
78					C								C				C															
76					C																											
74					F	R							F				A															F
72					F	R							F							R												
70																	F															
68					F	R							F				C		R													
67						R							A				A						R				R					C
65													F				A															
62					R												C									R						
61													F				F															
60				R	F	R																										R
59						F																										
58						C											A															
56					A	C											A															F
55					R	R																										
53					F	F											C		F													C
52					F	F											R														R	
51					C							R					A															R
50					C	F							F				R															
49					A	C							C																	R		F
46				F	C	C																									R	
44				C	A	A											A										F	R		F	R	
37				F	F	R																									R	R
34				A		F						F					A													R		
33																										R						
30				F											F													R				C
29																																R
28				C											A	R	R										R			R		
24											R				F												R					F
23				A							C																			R		
22											R																					
20										C	F																					
18										F																						
17														F					R													F
16					R																				R							
15							C	A	A					F																	R	C
10			F		F																									R		
8							R	A		R								R											R	F	F	A
6								R										R														R
1	R	F					C	R	A									A			R	F		F	R	R					F	C

Fig. 12. Species distribution and relative abundance for Section F, Las Vegas Range; locality 5247. Abbreviations: R=1, F=2–4, C=5–10, A=>10 specimens respectively.

Bed Number \ species	Cavusgnathus altus	C. naviculus	C. convexus	C. spathus	C. gigantus	C. lautus	Gnathodus commutatus	G. bilineatus	G. girtyi simplex	G. defectus	G. muricatus	G. opimus	G. bassleri	Streptognathodus unicornis	S. noduliferus	Idiognathodus delicatus	Spathognathodus campbelli	S. rexroadi	S. minutus	Neoprioniodus singularis	N? expandofundus	N. loxus	N. cf. camurus	Ozarkodina roundyi	O. delicatula	O. campbelli	Synprioniodina denticamura	Roundya sp.	Metalonchodina sp.	Ligonodina spp.	Lonchodina spp.	Hindeodella spp.
91					F								F			F																
89					F								A			F			F													
87					C	A							C			C																F
84					F	R																										
82					C	R							A			C			R											F	F	
80					R	F							A			A																
78					F								F			R																
74					C	F							A			A							R							R		
73													C			A																F
71													F			A																F
69					A	C							A			A															R	F
67a					F	F							A			A			R		R									R		C
67													C			C																R
65					C	F							C			F																
63					F	F						A	F			A						R										R
62					F	R										R																R
60					R	F										C															R	
58					C	C										F																
56					A	A										A																R
55					F	C							R			C			R											R		
53					C	C										F																
50				R		R																										
48				C	A	C									A	A															F	
47				F	A	A										C															R	
44				R											C											R						F
42					R								F		A	A																C
41				F	R										C											R						C
40					F					A	A																					
39					F	C					A																				F	
37					R	R				A	F																					
36											A																					
34			R		R			A	A		R			F																		C
32								R	R		R			R																		R
30			F								R			F																		F
26		F	F						F		F			C					R													F
24		F							C					C																		C
23											R			C					R											R		C
19														F																		
17								C	C																							R
16														F																	R	
13							F	A												C				F	F							A
11							F	A												R					C				F	C	A	A
7							F	C						F														R		R		C
4	A							R	F																							
1	A						C		A								A	C	R	F				F			F					A

Fig. 13. Species distribution and relative abundance for Section G, Dry Lake; locality 5248. Abbreviations: R=1, F=2–4, C=5–10, A=>10 specimens respectively.

Bed Number \ species	*Cavusgnathus altus*	*C. naviculus*	*C. convexus*	*C. spathus*	*C. giganteus*	*C. lautus*	*Gnathodus commutatus*	*G. bilineatus*	*G. girtyi simplex*	*G. muricatus*	*G. bassleri*	*Streptognathodus unicornis*	*S. expansus*	*S. noduliferus*	*Idiognathoides convexa*	*Idiognathodus delicatus*	*Spathognathodus minutus*	*S. coloradoensis*	*Neoprioniodus scitulus*	*N.? expandofundus*	*N. loxus*	*N. cf. camurus*	*Ozarkodina curvata*	*O. delicatula*	*O. campbelli*	*Synprioniodina denticamura*	*S. microdenta*	*Hibbardella ortha*	*H. sp. A.*	*Ligonodina spp.*	*Lonchodina spp.*	*Hindeodella spp.*
67						R																										
66						F										F	R										R					F
65					C	F										R					R				F							
64											R					A									R							
63					C	C					R					R	F												R			F
62						R																										
61					F	R										F		R														F
59					C	A					C					C																F
54					C	C																										
53					F	R					F					F																R
52					F	R					A					A	F			R					F							R
51					F	R					C					A		F							F							F
49					A	C					A					A														R	F	R
47						C																										
46					F	F										C	R															C
45					F	C										A									R					R		F
43					A	F										F	R				F											
42					A	A										R																F
40				R	R	R																										
39				C		C					R					C																
37				R	A	C											R				R	R										C
36				F	C	F																									R	
35					R																											
34				R	F	F								R		F												R				
30					R	R							R			R																
29					R																											F
28						R							R			A																F
27					F								F	A	R	A																
26				F										R																		R
25				R										A	C						F			R				R			F	F
24														F																		R
23					R					F				A	R															R	F	R
22					C	F				A					F						R									F	R	R
19										R																						
18					R				C	A																						F
16												F																				
15		F							C			A							R			R		F		R						
14								R	F			R										R	R									A
12		F			A							F																				
10								F	F			C					F					R										A
8								A																						F	R	F
7		C					R	A															R									F
4			R														F					R										C
1	A						R		A												R										F	C

Fig. 14. Species distribution and relative abundance for Section I, Apex; locality 5274. Abbreviations: R=1, F=2–4, C=5–10, A=>10 specimens respectively.

Bed Number \ species	*Magnilaterella robusta*	*Cavusgnathus altus*	*C. spathus*	*C. gigantus*	*C. lautus*	*Gnathodus commutatus*	*G. bilineatus*	*G. girtyi simplex*	*G. defectus*	*G. muricatus*	*G. opimus*	*G. bassleri*	*Streptognathodus unicornis*	*S. expansus*	*S. noduliferus*	*Idiognathoides convexa*	*Idiognathodus delicatus*	*Spathognathodus campbelli*	*S. rexroadi*	*S. minutus*	*S. coloradoensis*	*Neoprioniodus singularis*	*N. loxus*	*N. cf. camurus*	*Ozarkodina roundyi*	*O. curvata*	*O. delicatula*	*Synprioniodina denticamura*	*Hibbardella ortha*	*Roundya* sp.	*Metalonchodina* sp	*Ligonodina* spp.	*Lonchodina* spp.	*Hindeodella* spp.
69				C	C																			R										R
67				F	R							F								R												R		F
57				F	R							F																						
55												F					F																	R
53				F	C							A					A				R													C
51				R								R					C																	
50			C		C						F	R																						F
49				R	R																													
48												C					A																	
46				F	A												F																	
43				F	R																													
39				R																														
37			F												R		C																	
35				F								F																						
33					F																													
32					R												C																	
30																	C																	F
28															F	F																F		
27														F	A																	R		F
25															A								R				C					R		A
24				A	F					A													R									C	R	R
23				R					A	A																								
21				R	F			F		A																								
20							A	A																									F	
18								R																										R
16								F	R	R			C																				R	
11						R	A																				F			R		R	F	C
9							C													R														A
6						F	A	C					C																		F	R	F	F
1	R	A				C	R	A										A	F	A		C	R		F	F	F	F	R					A

Fig. 15. Species distribution and relative abundance for Section K, Gass Peak; locality 5277. Abbreviations: R=1, F=2–4, C=5–10, A=>10 specimens respectively.

The fauna from the top of the Battleship Wash formation is one of the richest and most diverse found. The association of *Cavusgnathus altus, Gnathodus commutatus, G. bilineatus, G. girtyi simplex, Neoprioniodus singularis, Ozarkodina roundyi, O. delicatula, Magnilaterella robusta, Spathognathodus campbelli,* and *Geniculatus*? sp. in this fauna indicates an early Chester age for the fauna. The latter three taxa were not found above this horizon. *Magnilaterella robusta* was previously known only from Meramec and lower Chester rocks of the Illinois Basin.

Faunas from the Indian Springs formation and basal part of the Bird Spring formation are dominated by *Gnathodus commutatus, G. bilineatus, G. girtyi sim-*

plex, Cavusgnathus naviculus, C. convexus, and *Streptognathodus unicornis.* In addition, the lower half of the Indian Springs formation contains *Neoprioniodus singularis* and *N. scitulus.* This association suggests correlation with the Glen Dean through Kinkaid formations of the Illinois Basin. The appearance of *Cavusgnathus spathus, C. gigantus,* and *C. lautus* in the upper part of the Indian Springs formation or the basal Bird Spring formation, associated with typical Chester gnathodids, suggests transition into the Morrow faunas where these species are common and typical Chester species are not present. The fauna described by Dunn (1965) is from this interval at Lee Canyon in the Spring Mountains. Therefore, the upper half of the Indian Springs and basal unit of the Bird Spring contain conodont faunas that are assigned a late Chester age, probably equivalent to the *Homoceras* Zone of Europe. Equivalent conodonts have not been recognized in the mid-continent although they may be present in the basal Imo formation of Arkansas.

Morrow and Derry faunas contain abundant *Cavusgnathus gigantus* and *C. lautus* throughout, but other species are more restricted. *Cavusgnathus spathus,* which appeared in latest Chester time, was found nearly to the top of the Morrow and although not abundant may be a good Morrow index fossil for southern Nevada. The base of the Pennsylvanian is marked by the appearance of *Streptognathodus noduliferus* and *Idiognathoides convexus,* two species described from the Dimple limestone of Morrow age of west Texas. These two forms were found to range into the late and middle Morrow time, respectively, in southern Nevada.

Three species of *Streptognathodus, S. lanceolatus,* n. sp., *S.* cf. *S. gracilis,* and *S. expansus* n. sp., appear to be restricted to the middle Morrow part of the Bird Spring. Four other species first appear in the middle Morrow: *Idiognathodus delicatus,* which is very abundant and continues through the Derry; *Ozarkodina campbelli,* which continues nearly to the top of the Derry; *Gnathodus bassleri,* which becomes very common in late Morrow time and extends through the Derry; and *G. opimus,* a species that continues into the *Fusulinella* Zone. This is the first report of *G. opimus* in North America; the species was originally described from Morrow rocks of Japan.

There is little change between Morrow and Derry faunas. Except for the disappearance of *Cavusgnathus spathus,* late Morrow species continued into Derry time. Three species, however, first appear in Derry time: *Idiognathodus claviformis* was found in abundance in the lower part of the *Profusulinella* Zone at Arrow Canyon; *Neoprioniodus? expandofundus* n. sp. was not found outside the *Profusulinella* Zone; and *Spathognathodus coloradoensis* appeared in the *Profusulinella* Zone and continued through the *Fusulinella* Zone.

ZONATION AND CORRELATION

Faunizones based on cephalopods, fusulines, and brachiopods have been recognized in the Late Mississippian and Pennsylvanian rocks of southern Nevada by different authors. Longwell and Dunbar (1936), followed by other authors, recognized the zone of *Fusulinella* within their second member of the Bird Spring formation. The *Profusulinella* Zone was reported at Lee Canyon (Rich, 1961) and Arrow Canyon (Cassity and Langenheim, 1966). The *Rhipidomella nevadensis* Zone (Sadlick, 1957) was recognized by Brill (1963), Lane (1963), Coogan (1964), and Langenheim and Langenheim (1965). Coogan also recorded the *Eumorphoceras* Zone in the upper part of his Chester sequence (units 1 and 2 of the Indian Springs herein). Bissell (1962) noted the presence of four fusuline zones within his Springer through Derry sequence.

Because of the abundance and diversity of fossils within the Late Carboniferous rocks of southern Nevada and the overlapping ranges of many species, several zones are believed applicable. Two new ones, the *Gnathodus girtyi simplex* and *Streptognathodus noduliferus—Idiognathoides convexus* Assemblage Zones,

System	Series	Midcontinent Zones: Fusulines Thompson (1948)	Midcontinent Zones: Goniatites Gordon (1965)	Southern Nevada Zones: Goniatites	Southern Nevada Zones: Fusulines	Southern Nevada Zones: Brachiopods	Southern Nevada Zones: Conodonts	Southern Nevada: Formations
Pennsylvanian	Derry	*Fusulinella*	*Pseudoparalegoceras kesslerense*		*Fusulinella-Chaetetes*			Bird Spring
Pennsylvanian	Derry	*Profusulinella*	*Pseudoparalegoceras kesslerense*	?	*Profusulinella*			Bird Spring
Pennsylvanian	Morrow	*Millerella*	*Gastrioceras branneri*	*Gastrioceras* sp. ? — ?	*Millerella*		*Streptognathodus noduliferus Idiognathoides convexa*	Bird Spring
Mississippian	Chester	?	? *Eumorphoceras bisulcatum*	? *Eumorphoceras bisulcatum*	?	*Rhipidomella nevadensis*	*Gnathodus girtyi simplex*	Indian Springs
Mississippian	Chester							Chainman shale / Battleship Wash

Fig. 16. Correlation of fossil zones recognized in Chester through Derry rocks of southern Nevada with selected zones of the midcontinent.

based on conodonts are proposed. The zone of *Gastrioceras,* which is recognized in the midcontinent, is reported but not defined. The relationships of these zones with the formations recognized, with each other, and with zones of the standard midcontinent area are illustrated in figure 16.

The base of the *Eumorphoceras bisulcatum* Zone has not been determined in southern Nevada, although it is within the Chainman shale at Kane Springs Wash and probably within the Battleship Wash formation to the south of Kane Springs Wash. The top is within the upper shale unit of the Indian Springs and contains the *Syngastrioceras walkeri* fauna.

Strata equivalent in age to the *Gastrioceras* Zone are known in the area of study but because specimens of the genus have not been found therein, the zone of *Gastrioceras* is included in figure 6 with question. A cephalopod fauna containing *Gastrioceras, Bisatoceras,* and *Branneroceras,* among others, was recently found at Indian Springs, Nevada, adjacent to the study area by N. Gary Lane (personal communication, 1966). Discovery of this fauna allows recognition of the *Gastrioceras* Zone in southern Nevada, but additional work is needed to determine the exact limits of the zone and specific identifications of the fossils.

Foraminiferal zones are directly comparable between southern Nevada and the midcontinent. No modification of these zones is believed necessary. The abundance and interbedded association of *Fusulinella* and *Chaetetes,* however, coupled with the larger size of *Chaetetes* (therefore, more easily located in the field) suggests that recognition of a "local" *Fusulinella-Chaetetes* Faunizone in southern Nevada is desirable. It should be emphasized that occasional coralla of *Chaetetes* are found in the basal Desmoines as shown in plate 1, sections B (Kane Springs Wash West), F (Las Vegas Range), and H (Arrowlime) and have been noted by Brill (1963) in the Spring Mountains.

The *Rhipidomella nevadensis* Faunizone is widely recognized in the Great Basin. The top of this zone corresponds with the Mississippian-Pennsylvanian boundary, at least in southern Nevada and possibly throughout the Great Basin. Although other brachiopod zones are not now recognized within the area, further study of the numerous brachiopods collected would probably permit definition of additional zones.

Two new zones are proposed, based on conodont faunas. The lower zone contains a rather diverse fauna (pl. 1) and is chronologically equivalent to the *Rhipidomella nevadensis* Zone. This is given the name of *Gnathodus girtyi simplex* Assemblage Zone because the subspecies is distinct and occurs in abundance throughout the zone. Few specimens of *G. girtyi simplex* were found below the zone in the Battleship Wash formation. Other conodont species characteristic of the assemblage zone include *Cavusgnathus naviculus, C. convexus, Gnathodus commutatus, G. bilineatus, G. defectus,* and *Streptognathodus unicornis.*

An upper *Streptognathodus noduliferus–Idiognathoides convexus* Assemblage Zone is proposed for the lower and middle Morrow part of the section. This corresponds to the range of *S. noduliferus* in southern Nevada. *Streptognathodus noduliferus* occurs throughout the zone and is abundant except in the basal and uppermost parts. *Idiognathoides convexus* is not abundant and was not found in the uppermost portion of the zone. This zone may be subdivided into three parts: a lower subzone containing abundant *Gnathodus muricatus* in addition to the name givers; an upper subzone with a few specimens of *Gnathodus bassleri* and numerous *Idiognathodus delicatus;* and a middle subzone lacking either of the supplemental elements of the lower and upper subzones.

ENVIRONMENTAL AND PALEOECOLOGIC IMPLICATIONS

Presence of the Las Vegas Hinge Line in southern Nevada is reflected in sediments of most Paleozoic systems by thinner shelf deposits east of the line and thicker basinal deposits west of the line. Chester through Derry sediments show the active influence of this line.

Brill (1963, fig. 2) showed the paleogeologic surface of the western Colorado Plateau and eastern Great Basin regions in early Chester time, which he called pre-Pennsylvanian, before deposition of the transition beds (Indian Springs formation herein). The line representing the eastern limit of Upper Mississipian limestone deposition (Battleship Wash–Bluepoint formations) extending northeast from Las Vegas into Utah approximates the Las Vegas Hinge Line of Welsh (1959). An abrupt change in lithology occurs above the Upper Mississippian limestones with deposition of the Indian Springs formation followed by the Bird Spring formation, sections C (Meadow Valley) through K (Gass Peak). To the northwest, Upper Mississippian dark shales (Chainman shale) were overlain by the Scotty Wash quartzite in the Pioche area and the Indian Springs formation in Kane Springs Wash, sections A and B. The abrupt change from carbonates of the Battleship Wash limestone to clastics in the Indian Springs suggests that a slight hiatus separated the two. If existent, this hiatus probably was short because no disconformable surface was developed nor do faunas show any significant change across this boundary. At Kane Springs Wash the lithologies are essentially the same on both sides of the time horizon and no hiatus is recognized.

The intercalated, thin-bedded shale, limestone, orthoquartzite, and rare conglomerate lithologies of the Indian Springs formation suggest a rapidly fluctuating environment of deposition. The even-bedded, fine-grained character of the sediments and scarcity of channeling implies a low-energy environment. The repetition of faunal elements in the same or similar lithologies laterally and vertically suggests an areal uniformity of differing environments. The abundance and diversity of the fauna would suggest a warm, shallow-water environment. The scarcity, and fragmented nature, of those few plant fossils present would imply noncontinental sedimentation; the presence of plant fossils throughout the area, however, probably implies a moderate proximity to land areas. Coogan (1964) considered this sequence at Arrow Canyon to represent tidal-flat and channel or nearshore deposits. Evidence listed above would tend to agree with his interpretation. A minor, area-wide hiatus is postulated for formation of the "marker" conglomerate.

Limestones of the Bird Spring formation, except for the southern facies of unit 2, are fine- to medium-grained, well-oxidized calcarenites and probably represent a moderate- to high-energy environment. The presence of fine-grained arenaceous material, abundance of disarticulated crinoidal remains, and presence of cross-bedding in many of the limestones would further support the interpretation of a high-energy environment. Periods of low energy are represented by argillaceous limestones and shales and typically contain a well-preserved diverse and abundant fauna. Development of widespread biostromal deposits suggests that conditions were fairly uniform over wide areas.

The southern facies of unit 2 is a fine-grained calcarenite to micrite, high in bituminous content. These beds contain a well-preserved diverse fauna with many delicate elements such as fenestrate bryozoans, chonetids, and thin-shelled productids with spines still attached and the specimens buried in living position. Abundance and diversity of the fauna suggests well-aerated water conditions; the high bituminous content and in-place preservaton imply a rapid rate of sedimentation.

SYSTEMATIC PALEONTOLOGY

All type specimens bear five digit catalog numbers (listed specimens four digit locality numbers) of the Invertebrate Paleontology Collections, Department of Geology, University of California, Los Angeles.

ORDER CONODONTOPHORIDA

Limestone and shale samples from sections B, D, F, G, I, and K were processed for conodonts. Except for one unidentified fragment all shale samples (84) were found to be barren of conodonts. Individual weight of limestone samples varied from 212 to 6,381 grams, with most samples weighing between 1,000 to 1,200 grams. A total of 325 limestone samples weighing more than 400 kilograms (23 samples unweighed) were dissolved in 12 percent acetic acid. Quartz-rich residues were concentrated using tetrabromoethane in heavy mineral separation techniques; iron-bearing residues were then run through a Frantz Isodynamic Magnetic Separator to remove the ferruginous minerals. The resulting residues were then picked and sorted. Identifiable conodonts were found in 269 samples, a success ratio of 83 percent. Abundancies ranged from 1 to 243 specimens per kilogram, averaging 27 specimens per kilogram.

Morphologic terms follow the usage of Hass (1962) with the exception that orientation of platform types is reversed to correspond with that advocated by Scott and others (1962).

Genus CAVUSGNATHUS Harris and Hollingsworth, 1933

Type species.—*Cavusgnathus altus* Harris and Hollingsworth, 1933, original designation.

Remarks.—Lane (1967) erected *Adetognathus* to include the forms, previously considered species of *Cavusgnathus,* which have a longer free blade than fixed blade. He referred all described Pennsylvanian species of *Cavusgnathus* to *Adetognathus.* This restricted *Cavusgnathus* to described species from the Mississippian, all of which are right forms. Some species referred to *Adetognathus* are left forms and others right forms. *Cavusgnathus spathus* Dunn (1966, p. 1297) would also be referred to *Adetognathus* (Lane, personal communication, March, 1967); although it contains both right forms and left forms.

Clearly there is some variation in ratio of length of free to fixed blade. Stratigraphically younger species tend to have a slightly longer free blade, with a mixing of the forms in Late Mississippian time. This suggests that the development of a longer free blade is an evolutionary trend for *Cavusgnathus* but it is not here considered to be of generic significance. *Adetognathus* is thus believed to be a synonym of *Cavusgnathus.* The problem of right form and left form is unsolved and is further complicated by the presence of a species (*C. spathus,* 125 specimens studied herein) containing both forms. This problem must await further study.

Cavusgnathus altus Harris and Hollingsworth
(Pl. 4, fig. 2)

1933 *Cavusgnathus alta* Harris and Hollingsworth, p. 201, pl. 1, figs. 10a, 10b.
1941*a* *Cavusgnathus cristata* Branson and Mehl, p. 177, pl. v, figs. 26–31.
1953 *Cavusgnathus cristata* Branson and Mehl. Hass, p. 77, pl. 14, figs. 12–14.
1956 *Cavusgnathus cristata* Branson and Mehl. Elias, p. 115, pl. ii, figs. 1–6.
1956 *Cavusgnathus cristata* var. *grandis* Elias, p. 115, pl. ii, figs. 12–14

1957 [Not] *Cavusgnathus cristata* Branson and Mehl. Bischoff, p. 19, pl. 2, figs. 7a, 7b.
1958 *Cavusgnathus cristata* Branson and Mehl. Rexroad, p. 16, pl. 1, figs. 15–17.
1961 *Cavusgnathus cristata* Branson and Mehl. Rexroad and Burton, p. 1151, pl. 138, fig. 16.
1966 *Cavusgnathus alta* Harris and Hollingsworth. Rexroad and Lane, p. 1391, text fig. 1.

Description.—See Harris and Hollingsworth (1933, p. 201); Branson and Mehl (1941*a*, p. 177).

Remarks.—Specimens of *Cavusgnathus altus* closely match the illustrations of *C. cristata* from Branson and Mehl (1941*a*, pl. v, figs. 26–31) and of the refigured holotype (Rexroad and Lane, 1966, p. 1391). This species is most abundant in the Battleship Wash formation, but the lower range limit was not determined. Specimens were not found above the lower part of the Indian Springs formation. Thus in southern Nevada *C. altus* may range through Chester time as has been reported for it in the midcontinent by other workers.

Material.—133 specimens.

Type.—Hypotype no. 43457.

Cavusgnathus convexus Rexroad

(Pl. 4, fig. 10)

1957 *Cavusgnathus convexa* Rexroad, p. 17, pl. 1, figs. 3–6.
1958 *Cavusgnathus convexa* Rexroad, p. 16, pl. 1, figs. 12–14.
1961 *Cavusgnathus convexa* Rexroad. Rexroad and Burton, p. 1151, pl. 138, fig. 14.
1964 *Cavusgnathus convexa* Rexroad. Rexroad and Furnish, p. 670, pl. 111, fig. 1.

Description.—See Rexroad (1957, p. 17).

Remarks.—Rexroad (1957) named *C. convexus*. Subsequently Rexroad and Collinson (1961, p. 5) stated that *C. convexus* may be a junior synonym of *C. altus* which was disclaimed by Rexroad and Lane (1966). Specimens from this study are all right forms, not abundant, and typically fragmentary; the few complete specimens compare closely with the original description and illustrations.

Material.—43 specimens.

Type.—Hypotype no. 43456.

Cavusgnathus gigantus Gunnell

(Pl. 4, fig. 6)

1933 *Cavusgnathus gigantus* Gunnell, p. 286, pl. 33, figs. 7, 8.
1933 *?Cavusgnathus missouriensis* Gunnell, pp. 286–287, pl. 33, figs. 10, 11.
1941 *Cavusgnathus giganta* Gunnell. Ellison, p. 126, pl. 21, figs. 44, 45, 49.
1941 *Cavusgnathus giganta* Gunnell. Ellison and Graves, no. 3, pl. 3, fig. 3.
1944 *Cavusgnathus giganta* Gunnell. Ellison in Branson, p. 325, pl. 45, figs. 44, 45, 49.
1949 *Cavusgnathus giganta* Gunnell. Youngquist and Downs, p. 162, pl. 30, figs. 18–20.
1949 *Cavusgnathus arca* Sturgeon and Youngquist, p. 383, pl. 75, figs. 11, 12.
1949 *Cavusgnathus unicornis* Youngquist and Miller, p. 619, pl. 101, figs. 18–23.
1952 *Cavusgnathus giganta* Gunnell. McLaughlin, p. 620, pl. 83, figs. 3, 4, 6, 7.
1957 *Cavusgnathus arca* Sturgeon and Youngquist. Bischoff, p. 19, table 2, figs. 10*a*, 10*b*.
1957 *Cavusgnathus unicornis* Youngquist and Miller. Rexroad, p. 17, pl. 1, fig. 7.
1957 *Cavusgnathus unicornis* Youngquist and Miller. Lys and Serre, p. 1042, pl. ii, figs. 3*a*, 3*b*.
1958 *Cavusgnathus unicornis* Youngquist and Miller. Rexroad, p. 17, pl. 1, figs. 6–11.
1961 *Cavusgnathus unicornis* Youngquist and Miller. Rexroad and Burton, p. 1152, pl. 138, figs. 10–12.
1963 *Cavusgnathus unicornis* Youngquist and Miller. Rexroad and Collinson, p. 9, pl. 1, figs. 26, 27.
1964 *Cavusgnathus unicornis* Youngquist and Miller. Rexroad and Furnish, p. 670, pl. 111, fig. 3.

1967 *Adetognathus giganta* (Gunnell). Lane, p. 931, pl. 120, figs. 16, 18, 19; pl. 121, figs. 8, 12, 13, 16.

Description.—See Gunnell (1933, p. 7); Ellison (1941, p. 126).

Remarks.—*Cavusgnathus gigantus* is the most abundant species of this genus in the Bird Spring formation and was not found below the Indian Springs formation. All specimens referred to this species typically have a large posterior denticle on the blade. The most pronounced variation of this species is in the shape of the posterior end, which ranges from sharply pointed to slightly rounded. Gradational specimens from one extreme to the other are present in the sample. On some specimens a short, pointed, aborally-curved posterior prong extends beyond the juncture of the two parapets. There is slight to moderate anterior-posterior arching. Rexroad and Collinson (1961) discuss *C. unicornis* and related forms. They conclude (p. 6): "Thus, most authors have used a broad interpretation of *C. giganta,* and, if such is correct, both *C. unicornis* and *C. arca* should be considered synonyms of *C. giganta.* Only study of large numbers of specimens from a broad geographic and stratigraphic range can offer a final solution."

C. unicornis was named from the Mississippian Pella beds of Iowa (Youngquist and Miller, 1949) and has subsequently been reported from several formations in the Chester series of the Illinois Basin (Rexroad, 1957, 1958; Rexroad and Burton, 1961; Rexroad and Collinson, 1963) and the late Visean of the Asturia region in Spain (Lys and Serre, 1957). *C. gigantus* has been reported from strata of Pennsylvanian and Early Permian age from the midcontinent region of North America by several workers. *C. arcus* was originally named from the Allegheny series of Ohio (Sturgeon and Youngquist, 1949) and has been recorded from the cuIII Teilzone in Germany (Bischoff, 1957). In southern Nevada *C. gigantus* ranges from the upper Chester part of the Indian Springs formation to at least the Desmoines part of the Bird Spring formation with the highest extent of the range not determined.

Thus, on the basis of the geographic and stratigraphic distribution, and by comparison of descriptions and illustrations of *C. gigantus, C. arcus,* and *C. unicornis,* I accept the broad interpretation for this species and consider the latter two named species synonyms of the former.

All the illustrated specimens listed in the synonomy as well as all specimens of *C. gigantus* from this study are right forms. Lindström (1964, p. 124) reported a personal communication from Dr. C. Rexroad that the Late Mississippian cavusgnathids from the Illinois Basin are one-formed only. It is interesting to observe that the range of *C. lautus,* a left form, closely parallels that of *C. gigantus* with the exception that the latter occurs stratigraphically before the former. *C. lautus,* 455 specimens, and *C. gigantus,* 514 specimens, are the most abundant representatives of this genus from the Bird Spring formation. If *C. gigantus* was not known to have a slightly earlier geologic occurrence than *C. lautus,* the Late Mississippian to Early Permian range of these two species in southern Nevada might suggest that they are the right form and left form, respectively, of a conodont pair.

Material.—545 specimens.

Type.—Hypotype no. 43458.

Cavusgnathus lautus Gunnell
(Pl. 4, fig. 9)

1933 *Cavusgnathus lautus* Gunnell, p. 286, pl. 31, figs. 67, 68.
1933 [Not] *Cavusgnathus lautus* Gunnell, p. 286, pl. 33, fig. 9.
1941 *Cavusgnathus lauta* Gunnell. Ellison, p. 126, pl. 21, figs. 47, 48.
1941 *Cavusgnathus lauta* Gunnell. Ellison and Graves, pl. 3, fig. 2.
1967 *Adetognathus lauta* (Gunnell). Lane, p. 933, pl. 121, figs. 1–5, 7, 10, 11, 15, 17, 18.

Description.—See Gunnell (1933, p. 286); Ellison (1941, p. 126).

Remarks.—*Cavusgnathus lautus* is abundant in the Bird Spring formation above the *Rhipidomella nevadensis* Zone. Some variation is noted in the form of the posterior oral surface, which varies from sharply pointed to slightly rounded. *C. flexus,* described by Ellison (1941, p. 126), was distinguished from *C. lautus* by the rounded apron on the oral surface of the platform. None of the specimens of *C. lautus,* but some of *C. gigantus,* found in the Bird Spring formation have as widely flaring aprons as *C. flexus.* A gradation toward a widely flaring apron is noted in *C. lautus,* and additional specimens may show that a complete gradation exists between *C. lautus* and *C. flexus,* in which case the latter should be considered a junior synonym of the former.

Material.—455 specimens.

Type.—Hypotype no. 43459.

Cavusgnathus naviculus (Hinde)
(Pl. 4, fig. 3)

1900 *Polygnathus navicula* Hinde, p. 342, pl. ix, fig. 5.
1928 *Polygnathus navicula* Hinde. Holmes, p. 18, pl. 7, fig. 14.
1947 *Cavusgnathus cristata* Branson and Mehl. Cooper, p. 91, pl. 20, figs. 4–10.
1957 *Cavusgnathus cristata* Branson and Mehl. Bischoff, p. 19, pl. 2, figs. 7*a*, 7*b*.
1960 *Cavusgnathus navicula* (Hinde). Clarke, p. 23, pl. iv, figs. 1–3.
1960 *Cavusgnathus inflexa* Clarke, p. 23, pl. iii, figs. 17–19.
1961 *Cavusgnathus navicula* (Hinde). Rexroad and Burton, pp. 1151–1152, pl. 139, figs. 4–13.
1961 *Cavusgnathus unicornis* Youngquist and Miller. Higgins, pl. x, fig. 3.
1965 *Cavusgnathus navicula* (Hinde). Dunn, p. 1147, pl. 140, figs. 10, 11.

Description.—See Hinde (1900, p. 342); Clarke (1960, p. 23).

Remarks.—All specimens identified as *Cavusgnathus naviculus* from this study are fragmentary; they closely match the illustrations of the holotype (refigured by Clarke, 1960, pl. iv, figs. 1–3). *C. naviculus* was found only in the Indian Springs formation.

Material.—19 specimens.

Type.—Hypotype no. 43465.

Cavusgnathus spathus Dunn
(Pl. 4, figs. 1, 4, 5)

1966 *Cavusgnathus spatha* Dunn, p. 1297, pl. 157, figs. 3, 7, 8.

Description.—See Dunn (1966, p. 1297).

Remarks.—*Cavusgnathus spathus* is a paired form, both right form and left form occurring in the same sample, which is not common among the cavusgnathids. The range of *C. spathus* in southern Nevada was found to extend from the uppermost part of the *Rhipidomella nevadensis* Zone to a short distance below the top of the Morrow series. Specimens are most abundant throughout the middle Morrow part of the Bird Spring formation. *C. spathus* was probably derived from *C. gigantus* in late Chester time by addition of the posterior extension. The deflection of the blade at the posterior end to merge with the blade parapet is somewhat suggestive of relation to *Streptognathodus;* the blade, however, is not positioned at the center of the trough as typical of that genus.

Material.—125 specimens.

Types.—Hypotypes nos. 43460–43464.

Genus GENICULATUS Hass, 1953

Type species.—*Polygnathus? claviger* Roundy, 1926, original designation.

Geniculatus? sp.
(Pl. 8, fig. 1)

Remarks.—A single fragmentary specimen consisting of three slightly appressed denticles on a wide tapering bar is identified as *Geniculatus?* sp. This specimen was found in the Battleship Wash formation and resembles part of the wide posterior bar of *G. claviger,* a species reported from the Barnett formation of Texas (Hass, 1953).

Type.—Hypotype no. 43455.

Genus GNATHODUS Pander, 1856

Type species.—*Gnathodus mosquensis* Pander, 1856, monotypy.

Gnathodus bassleri (Harris and Hollingsworth)
(Pl. 5, figs. 9, 14, 15)

1933 *Polygnathus bassleri* Harris and Hollingsworth, p. 198, pl. 1, fig. 13.
1933 *Polygnathus wapanuckensis* Harlton, p. 15, pl. 4, fig. 13.
1941 *Gnathodus wapanuckensis* (Harlton). Ellison and Graves, pl. 2, figs. 13–17.
1949 [?] *Streptognathodus* sp. Youngquist and Downs, pp. 170–171, pl. 13, figs. 8–10.
1967 *Gnathodus bassleri bassleri* (Harris and Hollingsworth). Lane, p. 935, pl. 120, figs. 1, 3–5, 9–12, 15; pl. 123, figs. 1–6.
1967 *Gnathodus bassleri symmetricus* Lane, p. 935, pl. 120, figs. 2, 13–14, 17; pl. 121, figs. 6, 9.
1967 *Gnathodus wapanuckensis* (Harlton). Koike, p. 300, pl. 1, figs. 22–25.

Description.—See Harris and Hollingsworth (1933, p. 198); and Lane, (1967, p. 934).

Remarks.—Lane (1967) noting that *Gnathodus bassleri symmetricus* is more symmetrical, has a more central carina, and possesses a slightly deeper basal cavity distinguished it from *Gnathodus bassleri bassleri.* Specimens referred to both forms are common in southern Nevada and only the larger gerontic forms would be referred to *G. bassleri bassleri,* while the smaller would be considered *G. bassleri symmetricus.* Because both forms occur together throughout the same stratigraphic range they are considered growth stages.

G. bassleri possibly evolved from *G. girtyi* or *G. defectus* in early Morrow time by development of pronounced lateral parapets. An alternate lineage may be development from one of the streptognathodids such as *S. gracilis* by widening of the platform and posterior extension of the main carina.

The upper extent of the range in southern Nevada of *G. bassleri* is not determined. The species is not found below the lower middle Morrow part of the

Bird Spring formation, however, and it is not abundant until late Morrow time. The species is abundant throughout Derry time.

Material.—498 specimens.

Types.—Hypotypes nos. 43478–43480.

Gnathodus bilineatus (Roundy)

(Pl. 5, figs. 11, 12)

1926 *Polygnathus bilineatus* Roundy, p. 13, pl. iii, figs. 10a–10c.
1926 *Polygnathus texanus* Roundy, p. 14, pl. iii, figs. 13a, 13b.
1939 *Gnathodus bilineatus* Roundy. Cooper, p. 388, pl. 42, figs. 59, 60 (figures may not be).
1949 [?] *Gnathodus liratus* Youngquist and Miller, pp. 619–620, pl. 101, figs. 15–17.
1951 *Gnathodus bilineatus* (Roundy). Hass, pl. 1, fig. 1.
1956 [Not] *Gnathodus bilineatus* (Roundy). Bischoff and Ziegler, p. 146, pl. 11, figs. 21–23
1956 *Gnathodus pustulosus* Branson and Mehl. Elias, p. 115, pl. 3, figs. 1–8.
1956 *Gnathodus bilineatus* (Roundy). Elias, p. 118, pl. 3, figs. 23–29.
1956 *Gnathodus* cf. *bilineatus* Elias, p. 118, pl. 3, fig. 40.
1956 *Gnathodus streptognathoides* Elias, p. 119, pl. 3, figs. 54–57.
1957 *Gnathodus bilineatus bilineatus* (Roundy). Bischoff, pp. 21–22, pl. 3, figs. 11, 15–20; pl. 4, fig. 1.
1957 *Gnathodus bilineatus bilineatus* (Roundy). Flügel and Ziegler (part), pp. 38–39, pl. 3, figs. 1, 2 only.
1957 *Gnathodus modocensis* Rexroad, pp. 30–31, pl. 1, figs. 15–17.
1958 *Gnathodus bilineatus* (Roundy). Stanley, pp. 464–465, pl. 68, fig. 7.
1958 *Gnathodus modocensis* Rexroad, pp. 17–18, pl. 1, figs. 1, 2.
1958 *Gnathodus bilineatus bilineatus* (Roundy). Lys and Serre, p. 891, pl. ix, figs. 1a, 1b.
1959 *Gnathodus bilineatus* (Roundy). Voges, pp. 282–283, pl. 33, figs. 28–38.
1959 *Gnathodus* (*Harltonodus*) *bilineatus* (Roundy). Elias, pp. 145–146, pl. 1, figs. 3–12.
1959 *Gnathodus* (*Harltonodus*) *bransoni* Elias, pp. 147–148, pl. 1, figs. 13–18.
1959 [?] *Gnathodus* (*Harltonodus?*) *liratus* Youngquist and Miller. Elias, p. 148, pl. 1, figs. 19–21.
1959 *Gnathodus* (*Harltonodus*) *minutus* Elias, pp. 148–149, pl. 1, figs. 22–25.
1959 *Gnathodus* (*Harltonodus*) *multilineatus* Elias, p. 149, pl. 1, figs. 26–28.
1960 *Gnathodus smithi* Clarke, pp. 26–27, pl. iv, figs. 13, 14; pl. v, figs 9, 10.
1961 *Gnathodus bilineatus* (Roundy). Higgins, pl. x, fig. 5.
1962 *Gnathodus bilineatus* (Roundy). Higgins, pl. 2, fig. 25; pl. 3, fig. 32.
1962 *Gnathodus bilineatus bilineatus* (Roundy). Meischner, fig. 10.
1963 *Gnathodus bilineatus bilineatus* (Roundy). Bouckaert and Higgins, fig. 3.
1964 *Gnathodus bilineatus modocensis* Rexroad. Rexroad and Furnish, pp. 670–671, pl. 111, figs. 4–5.
1965 *Gnathodus bilineatus* (Roundy). Dunn, p. 1148, pl. 140, figs. 7–9.
1966 *Gnathodus* (*Harltonodus*) *bilineatus* (Roundy). Elias, p. 11, pl. 1, figs. 3–12.
1966 *Gnathodus* (*Harltonodus*) *bilineatus smithi* Clarke. Elias, p. 12, pl.1, figs. 13–14.
1966 *Gnathodus* (*Harltonodus*) *bransoni* Elias, p. 12, pl. 1, figs. 15–20.
1966 *Gnathodus* (*Harltonodus*) *delicatus hassi* Elias, p. 15, pl. 1, fig. 21.
1966 *Gnathodus* (*Harltonodus?*) *liratus* (Youngquist and Miller). Elias, p. 15, pl. 1, figs. 22–24.
1966 *Gnathodus* (*Harltonodus*) *minutus* Elias, p. 16, pl. 1, figs. 25–28.
1966 *Gnathodus* (*Harltonodus*) *multilineatus* Elias, p. 17, pl. 1, figs. 29–31.
1966 *Gnathodus* (*Harltonodus*) cf. *G.* (*H.*) *multilineatus* Elias, p. 17, pl. 1, fig. 32.
1967 *Gnathodus bilineatus* (Roundy). Globensky, p. 440, pl. 58, figs. 9, 13.
1967 *Gnathodus bilineatus* (Roundy). Koike, p. 296, pl. 1, figs. 9–11.

Description.—See Roundy (1926, p. 13); Hass (1953, p. 79).

Remarks.—*Gnathodus bilineatus* is one of the most abundant species present in the Indian Springs formation and basal part of the Bird Spring formation.

Considerable variation is shown by specimens within individual samples. The outline of the outer platform tends to be more triangular (widest anteriorly) in immature specimens becoming more rectangular in adult forms. The platform ornamentation varies from semialigned nodes to well-aligned nodes to ridges that on the vast majority of the specimens tend to curve toward the medial carina posteriorly. The number of ridges or rows of aligned nodes ranges from two to six. Most adult and gerontic specimens have a low ridge developed laterally along the anterior oral margin of the platform. As a result the species is considered to be variable in form regarding ornamentation and shape during growth. Rexroad and Collinson (1961, pp. 7–8) discuss the synonomy, occurrence, and subspeciation of *G. bilineatus.* Although in general agreement with their discussion, I consider *G. modocensis* a junior synonym and geographic variety of *G. bilineatus.*

Material.—783 specimens.

Types.—Hypotypes nos. 43466–43467.

Gnathodus commutatus (Branson and Mehl)

(Pl. 5, fig. 13)

1941*a* *Spathognathodus commutatus* Branson and Mehl, p. 172, pl. 5, figs. 19–22.
1941*b* *Spathognathodus commutatus* Branson and Mehl, p. 98, pl. 19, figs. 1–4.
1941 *Spathognathodus commutatus* Branson and Mehl. Ellison and Graves, p. 3, pl. 2, fig. 6.
1953 *Gnathodus inornatus* Hass, p. 80, pl. 14, figs. 9–11.
1956 *Spathognathodus commutatus* Branson and Mehl. Elias, p. 119, pl. iii, figs. 19–22.
1956 *Spathognathodus inornatus* (Hass). Elias, p. 119, pl. iii, figs. 37–39.
1956 *Spathognathodus* cf. *inornatus* (Hass). Elias, p. 119, pl. iii, figs. 41, 42, 62, 63.
1957 *Gnathodus commutatus commutatus* (Branson and Mehl). Bischoff, p. 23, pl. 4, figs. 2–6, 15.
1957 *Spathognathodus* cf. *commutatus* Branson and Mehl. Rexroad, p. 38, pl. 3, figs. 23, 24.
1957 *Gnathodus commutatus commutatus* (Branson and Mehl). Flügel and Ziegler, p. 39, pl. iii, fig. 21.
1958 *Gnathodus inornatus* Hass. Stanley, p. 465, pl. 68, figs. 5, 6.
1958 *Gnathodus commutatus commutatus* (Branson and Mehl). Lys and Serre, p. 891, pl. ix, figs. 2a, 2b.
1958 *Spathognathodus* cf. *S. commutatus* Branson and Mehl. Rexroad, p. 26, pl. 6, fig. 8.
1959 *Gnathodus commutatus commutatus* (Branson and Mehl). Voges, p. 281.
1960 *Spathognathodus commutatus* Branson and Mehl. Clarke, p. 19, pl. iii, figs. 4, 5.
1961 *Gnathodus commutatus commutatus* Bischoff. Higgins, pp. 212–213, pl. x, fig. 6, text fig. 6*a*.
1961 *Gnathodus commutatus* (Branson and Mehl). Rexroad and Burton, p. 1153, pl. 139, figs. 1–3.
1962 *Gnathodus commutatus commutatus* (Branson and Mehl). Higgins, pl. 2, fig. 22.
1962 *Gnathodus commutatus commutatus* (Branson and Mehl). Meischner, p. 30, fig. 10.
1963 *Gnathodus commutatus commutatus* (Branson and Mehl). Bouckaert and Higgins, fig. 3.
1967 *Gnathodus commutatus commutatus* (Branson and Mehl). Koike, p. 296, pl. 1, figs. 12–16.

Description.—See Branson and Mehl (1941), p. 172.

Remarks.—Rexroad and Collinson (1961, p. 6) discuss the generic assignment of *Gnathodus commutatus* and I agree with their conclusions. *G. commutatus* is not abundant in the samples studied, but specimens were found in several samples. The lower extent of the range of this species in southern Nevada was not determined. *G. commutatus* was not found above the *Rhipidomella nevadensis* Zone.

Material.—50 specimens.

Type.—Hypotype no. 43468.

Gnathodus defectus Dunn
(Pl. 5, fig. 16)

1966 *Gnathodus defectus* Dunn, p. 1300, pl. 158, figs. 1, 5.

Description.—See Dunn (1966, p. 1300).

Remarks.—In southern Nevada *Gnathodus defectus* was found in Late Mississipian age strata and ranges into lowermost Morrow strata. Dunn (1966) reported it only from the lower Morrow from the same area. It is most abundant in the upper part of the *Rhipidomella nevadensis* Zone and immediately above.

Material.—151 specimens.

Type.—Hypotype no. 43483.

Gnathodus girtyi simplex Dunn
(Pl. 5, fig. 10)

1965 *Gnathodus girtyi simplex* Dunn, p. 1148, pl. 140, figs. 2, 3, 12.

Description.—See Dunn (1965, p. 1148).

Remarks.—*Gnathodus girtyi simplex* is abundant from the first limestones in the Indian Springs formation to the top of the *Rhipidomella nevadensis* Zone in the base of the Bird Spring formation. Growth stages are especially well represented and show a marked uniformity of ornamentation and shape. The lower extent of the range of this species was not determined for southern Nevada.

Material.—787 specimens.

Type.—Hypotype no. 43469.

Gnathodus muricatus (Dunn, 1965), n. comb.
(Pl. 5, figs. 1–7)

1965 *Cavusgnathus muricata* Dunn, pp. 1147–1148, pl. 140, figs. 1a–c, 4.
1966 *Cavusgnathus transitoria* Dunn, p. 1299, pl. 157, figs. 9, 13.
1966 *Rhachistognathus prima* Dunn, p. 1301, pl. 157, figs. 1, 2.

Emended description.—In oral view unit narrow, elongate, lance-shaped; blade straight, laterally compressed, one-half to three-fifths unit length, continues onto platform as medial carina; platform variable, straight to gently bowed inward, pointed posteriorly, formed by medial carina, inner and outer parapets; inner parapet parallel to blade in anterior half, slightly convex outward in posterior half, composed of four to seven discrete pillar-like nodes, may merge with medial carina at posterior tip or at second or third node anterior to posterior tip; secondary trough between medial carina and inner parapet narrow, shallow to moderately deep; medial carina with two to five nodes at posterior tip, becoming a faint to nonexistent ridge through middle of platform or may be subdued along inner edge of nodes of outer parapet, posteriorly formed by solid ridge of three or four fused denticles; outer parapet very narrow, gently convex outward, composed of one to five pillar-like nodes which may be distinctly set out from medial carina by faint secondary trough or appear as slightly outward set secondary row of nodes along medial carina, gently merges with medial carina anteriorly, more abrupt merger posteriorly at second or third node of carina anterior to posterior tip.

In lateral view oral surface concave centrally; blade usually highest anteriorly descending to platform, may have two to four denticles anterior to highest which are progressively smaller anteriorly, composed of six to thirteen appressed denticles fused nearly to tips; denticles largest where highest and decrease in size with descending height; platform moderately high with posterior end arched aborally; aboral surface straight to convex aborally.

In aboral view pulp cavity smooth, moderately shallow, widest at anterior end of platform tapering to medial groove at posterior tip, continues as medial groove along blade, becoming less apparent anteriorly; aprons thin, delicate; outer apron evenly flared, widest centrally; inner apron widest anteriorly, sharply narrowing posteriorly.

Remarks.—The three specimens on which *Cavusgnathus muricatus,* the six specimens on which *Cavusgnathus transitoria,* and the twenty-nine specimens on which *Rhachistognathus prima* were based clearly belong to the same species and show part of the variation of the oral surface, especially when compared with the 613 specimens of this study. The abundant specimens of *Gnathodus muricatus* (Dunn) found during this investigation include growth stages from the same sample, showing the variable nature of the oral surface. The presence of a medial carina (often only weakly developed) and inner and outer parapets precludes assigning this species to *Cavusgnathus,* a genus typified by two parapets and a medial trough on the platform. *G. muricatus* probably developed from *G. girtyi simplex* in late Chester time. One transitional specimen identified as *G. girtyi simplex* (pl. 5, fig. 8) shows the outer parapet closely parallel to the faint medial carina and the development of the nodelike denticles on the posterior tip. Thus *G. muricatus* has the essential characteristics of *Gnathodus*—a medial carina meeting the platform in a nearly central position and containing a deep, expanded pulp cavity. *Rhachistognathus* is considered a junior synonym of *Gnathodus.*

A second possibility is that *G. muricatus* developed from *Spathognathodus* by addition of lateral parapets. Forms quite similar to *G. muricatus* have been reported from Devonian strata by different authors, and these have been assigned to *Spathognathodus.*

The stratigraphic range of *G. muricatus* in southern Nevada is restricted to late Chester, above the "marker" conglomerate, into the basal Morrow. Specimens are most common in the upper 20 feet of the Indian Springs and basal part of the Bird Spring.

Material.—613 specimens.

Types.—Hypotypes nos. 43470–43477.

Gnathodus opimus Igo and Koike
(Pl. 5, figs. 19–21)

1964 *Gnathodus opimus* Igo and Koike, pp. 189–190, pl. 28, figs. 15–18.
1965 [?] *Gnathodus opimus* Igo and Koike, p. 89, pl. 9, figs. 1–8.
1967 *Gnathodus opimus* Igo and Koike. Koike, p. 298, pl. 1, figs. 20–21.

Description.—See Igo and Koike (1964, pp. 189–190).

Remarks.—*Gnathodus opimus* was first reported from the Omi limestone of Early Pennsylvanian age from Japan (Igo and Koike, 1964). Discovery of this species in southern Nevada extends the geographic range to western North America. In southern Nevada *G. opimus* was found to range from uppermost Morrow into the middle part of the *Fusulinella* Zone.

Igo and Koike (1965) reported *G. opimus* from the Akiyoshi limestone. They indicated (p. 85) a probable Early Pennsylvanian age, but possible latest Chester age, for this limestone. Specimens from the Akiyoshi limestone show variation of the platform, with some forms having two parapets instead of one and less flaring of the basal cavity lips. This variation was not noted in the specimens studied here. Because specimens from the Akiyoshi limestone do not correspond with the original description and are stratigraphically older, they are doubtfully included in the synonomy.

Material.—28 specimens.

Types.—Hypotypes nos. 43481–43482.

Genus HIBBARDELLA Ulrich and Bassler in Bassler, 1925

Type species.—*Prioniodus angulatus* Hinde, 1879, original designation.

Hibbardella ortha Rexroad

(Pl. 5, fig. 22)

1958 *Hibbardella ortha* Rexroad, p. 18, pl. 2, figs. 9–12.
1960 *Hibbardella ortha* Rexroad. Clarke, p. 6, pl. 1, fig. 7.
1961 *Hibbardella ortha* Rexroad. Rexroad and Burton, pp. 1153–1154, pl. 140, figs. 5–6.
1964 *Hibbardella ortha* Rexroad. Rexroad and Furnish, p. 671, pl. 111, fig. 16.

Description.—See Rexroad (1958, p. 18).

Remarks.—The few specimens of *Hibbardella ortha* which are complete, except for the broken off posterior bar, closely compare with the original description and illustrations. Specimens were found in the Battleship Wash formation and Morrow part of the Bird Spring formation. Rexroad and Collinson (1961) reported a Middle and Late Mississippian range for *H. ortha* which should be extended to include Early Pennsylvanian.

Material.—6 specimens.

Type.—Hypotype no. 43529.

Hibbardella sp. A

(Pl. 8, fig. 14)

Description.—Anterior arch moderately deep, gently rounded; anterior face shallowly concave; lateral limbs project slightly anteriorly distally; anterior faces smooth, inclined 5° to 10° posteriorly aborally, each limb with three discrete, erect, peglike denticles, smaller distally; aboral margin with wide medial longitudinal groove extending to small pit at aboral end of cusp. Main cusp round to slightly compressed laterally, strongly arched posteriorly, tip broken off. Posterior bar heavy, wide, with three posteriorly arched, discrete, slightly laterally compressed denticles; tips of denticles not present; aboral surface wide median groove longitudinally extending anteriorly to small pit at aboral end of cusp.

Remarks.—A single specimen of *Hibbardella* sp. A was found in the *Fusulinella* Zone in Section I. It is easily distinguished from all other species of the genus by the coarse features and wide median grooves on the aboral surface of the two lateral and posterior bars. Because only one specimen was found and the posterior bar was broken during preparation, no specific name is assigned.

Material.—1 specimen.

Type.—Hypotype no. 43530.

Genus HINDEODELLA Ulrich and Bassler in Bassler, 1925

Type species.—*Hindeodella subtilis* Ulrich and Bassler in Bassler, 1925, original designation.

Hindeodella spp.
(Pl. 6, figs. 16–20)

Remarks.—Fragments of *Hindeodella* spp. were found in approximately 50 percent of the samples studied. Of nearly 700 specimens identified or referred to *Hindeodella* herein, less than 1 percent are relatively complete individuals. Because the genus is of insignificant stratigraphic use and because of the fragmentary condition of specimens found, no attempt was made at specific identification. A few individuals are illustrated to show a representation of forms found.

Material.—688 specimens.

Types.—Hypotypes nos. 43542–43546.

Genus IDIOGNATHODUS Gunnell, 1931

Type species.—*Idiognathodus claviformis* Gunnell, 1931, original designation.

Idiognathodus claviformis Gunnell
(Pl. 6, fig. 13)

1931 *Idiognathodus claviformis* Gunnell (part), p. 249, pl. 29, fig. 21 only.
1933 *Idiognathodus fustiformis* Gunnell, p. 269, pl. 31, fig. 7.
1933 *Idiognathodus nodostriatus* Gunnell, p. 272, pl. 31, fig. 20.
1941 *Idiognathodus claviformis* Gunnell. Ellison, p. 137, pl. 23, figs. 12–17, 19, 20, 22 (misprinted as figs. 12, 14–18, 20, 21, 23).

Description.—See Gunnell (1931, p. 249); Ellison (1941, p. 137).

Remarks.—The short, wide, posteriorly rounded and irregularly ornamented platform is diagnostic of *Idiognathodus claviformis.* Specimens studied herein show more regularity of transverse ridges, but variations were observed on individual specimens and the slight differences are not considered to be sufficient to justify erection of a new species. *I. claviformis* is known in the midcontinent from rocks of Desmoines through Virgil age. All specimens from this study are from the lower part of the *Profusulinella* Zone of Derry age, thus extending the stratigraphic and geographic range.

Material.—12 specimens.

Type.—Hypotype no. 43499.

Idiognathodus delicatus Gunnell
(Pl. 6, figs. 6–12)

1931 *Idiognathodus delicatus* Gunnell, p. 250, pl. 29, figs. 23–25.
1932 *Idiognathodus delicatus* Gunnell. Stauffer and Plummer, p. 45, pl. 4, figs. 4, 21, 24–26.
1932 *Idiognathodus magnificus* Stauffer and Plummer, p. 46, pl. 4, figs. 8, 18–20.
1933 *Idiognathodus pustulata* Harris and Hollingsworth, p. 204, pl. 1, fig. 11.
1933 *Idiognathodus cuneiformis* Gunnell, p. 270, pl. 31, fig. 8.
1933 *Idiognathodus harkeyi* Gunnell, p. 270, pl. 31, fig. 11.
1933 *Idiognathodus siculus* Gunnell, p. 271, pl. 31, fig. 14.
1933 *Idiognathodus modulatus* Gunnell, p. 271, pl. 31, fig. 15.
1933 *Idiognathodus sulciferus* Gunnell, p. 271, pl. 31, fig. 16.
1933 *Idiognathodus claratus* Gunnell, pp. 271–272, pl. 31, fig. 19.
1933 *Idiognathodus ruidus* Gunnell, p. 272, pl. 31, fig. 25.
1933 *Idiognathodus spathodus* Gunnell, p. 273, pl. 31, fig. 28.
1933 *Idiognathodus semipapulatus* Gunnell, p. 273, pl. 31, figs. 29, 50.

1933 *Idiognathodus megistus* Gunnell, p. 273, pl. 31, fig. 30.
1933 *Idiognathodus lanceolatus* Gunnell, pp. 273–274, pl. 31, figs. 31–32.
1933 *Idiognathodus folium* Gunnell, p. 274, pl. 31, fig. 33.
1933 *Idiognathodus cicatricosus* Gunnell, p. 274, pl. 31, fig. 34.
1933 *Idiognathodus wintersetensis* Gunnell, p. 274, pl. 31, figs. 36, 51.
1933 *Idiognathodus strigillatus* Gunnell, pp. 274–275, pl. 31, fig. 37; pl. 32, fig. 8.
1933 *Idiognathodus gemmiformis* Gunnell, p. 275, pl. 31, fig. 44.
1933 *Idiognathodus vadosus* Gunnell, p. 275, pl. 31, fig. 45.
1933 *Idiognathodus erodus* Gunnell, p. 275, pl. 31, fig. 48.
1933 *Idiognathodus fusiformis* Gunnell, p. 276, pl. 31, fig. 49.
1933 *Idiognathodus cherryvalensis* Gunnell, p. 277, pl. 32, fig. 5.
1933 *Idiognathodus corrugatus* Gunnell (part), p. 277, pl. 32, fig. 6 only.
1933 *Idiognathodus walteri* Gunnell, p. 277, pl. 32, fig. 9.
1933 *Idiognathodus warei* Gunnell, p. 279, pl. 32, figs. 59–61.
1933 *Idiognathodus kansensis* Gunnell, p. 279, pl. 32, figs. 62–64.
1941 *Idiognathodus delicatus* Gunnell. Ellison, pp. 134–135, pl. 22, figs. 31–36.
1941 *Idiognathodus magnificus* Stauffer and Plummer. Ellison, pp. 135–136, pl. 23, figs. 2, 3, 6, 9 (misprinted as 2, 4, 7, 10).
1941 *Idiognathodus acutus* Ellison, p. 137, pl. 23, figs. 21, 24 (misprinted as figs. 22, 25).
1941 *Idiognathodus delicatus* Gunnell. Ellison and Graves, pl. 3, figs. 20, 23.
1941 *Idiognathodus magnificus* Stauffer and Plummer. Ellison and Graves, pl. 3, figs. 25–27.
1944 *Idiognathodus delicatus* Gunnell. Branson and Mehl *in* Shimer and Shrock, p. 246, pl. 94, figs. 56–58.
1944 *Idiognathodus delicatus* Gunnell. Ellison *in* Branson, pl. 46, figs. 31–36.
1952 *Idiognathodus delicatus* Gunnell. McLaughlin, p. 619, pl. 83, figs. 8–11.
1952 *Idiognathodus magnificus* Stauffer and Plummer. McLaughlin, p. 619, pl. 83, figs. 12–14.
1960 *Idiognathodus* cf. *magnificus* Stauffer and Plummer. Clarke, p. 28, pl. 5, figs. 1–5.
1965 *Idiognathodus* cf. *I. magnificus* Stauffer and Plummer. Murray and Chronic, p. 601, pl. 71, figs. 7–12.
1965 *Idiognathodus meekerensis* Murray and Chronic, pp. 601–602, pl. 71, figs. 13–29.
1966 *Idiognathodus humerus* Dunn, p. 1300, pl. 158, figs. 6, 7.
1966 *Idiognathodus incurvus* Dunn, p. 1301, pl. 158, figs. 2, 3.
1966 *Streptognathodus angustus* Dunn, p. 1302, pl. 158, figs. 11–13.
1966 *Streptognathodus parvus* Dunn, p. 1302, pl. 158, figs. 9, 10.
1967 *Idiognathodus* sp. Lane, p. 936, pl. 119, figs. 10, 11.
1967 *Idiognathodus delicatus* Gunnell. Koike, p. 304, pl. 2, figs. 18–23.

Description.—See Gunnell (1931, p. 250); Ellison (1941, pp. 134–136).

Remarks.—*Idiognathodus delicatus* is the most abundant species found through the middle Morrow and Derry part of the Bird Spring formation. Growth stages show ontogenetic development and the numerous individuals show the wide variation of ornamentation for this species.

The platform of immature individuals consists of a median carina and two subparallel lateral parapets, all of subequal height and denticle size, which are terminated by mutual coalescence at the posterior end. Transverse ridges are developed on the posterior end in minute forms, and with continued growth ridges are added anteriorly ultimately covering the platform. Depending upon growth stage and variation the number of ridges ranges from one to twelve. Accessory lobes are developed simultaneously or independently; if independently, the larger inner lobe develops first and may appear before or after the anterior end of the platform is covered with transverse ridges. The accessory lobes may be set off by a narrow groove or merge smoothly into the transversely ridged portion of the platform.

Gradations between these two extremes were found, with the latter most common among larger or gerontic individuals indicating that the groove is filled as growth continues. Some variation occurs in the amount of inward bowing of the platform ranging from nearly straight to moderate. The wide variation of form of *I. delicatus* is reflected in the lengthy synonomy.

Ellison (1941) indicated the variable ornamentation of *I. magnificus* and distinguished it from *I. delicatus,* as the latter is smaller, more slender, and the accessory lobes are clearly set off from the main body of the platform. *I. magnificus* is here considered to be the gerontic form of *I. delicatus* as suggested by Ellison (p. 135) and is consequently a junior synonym. Murray and Chronic (1965) recognized *I. meekerensis* on the basis of four to seven transverse ridges instead of six to eighteen as reported by Ellison (1941) for *I. delicatus.* This is not of specific significance in view of the growth stages studied here.

Material.—1,724 specimens.

Types.—Hypotypes nos. 43500–43506.

Genus IDIOGNATHOIDES Harris and Hollingsworth, 1933

Type species.—*Idiognathoides sinuatus* Harris and Hollingsworth, 1933, original designation.

Remarks.—The conclusions of Lane (1967) after restudy of the type specimens of *Idiognathoides sinuatus* that they are not a cavusgnathid and that *Polygnathodella* is a junior synonym of *Idiognathoides* are here accepted. But inclusion of *Streptognathodus noduliferus* in this genus is not accepted. *S. noduliferus* has noded parapets (not transverse ridges, except on the posteriormost tip), a blade that adjoins the platform axially (not the outer parapet), and a short medial carina that distinctly stops or bends to merge into the outer parapet, leaving a medial trough on the posterior one-half to two-thirds of the platform.

Idiognathoides convexus Ellison and Graves
(Pl. 5, figs. 17, 18)

1941 *Polygnathodella convexa* Ellison and Graves, p. 9, pl. 3, figs. 10, 12, 16.
1941 *Cavusgnathus sinuata* Harris and Hollingsworth. Ellison and Graves (part), pl. 3, fig. 5 only.

Description.—See Ellison and Graves (1941, p. 9).

Remarks.—Specimens of *Idiognathoides convexus* show some variation in convexity, posterior sharpness, and development of an oral trough. Convexity of the oral surface ranges from straight to moderate; lateral convexity ranges from nearly straight to moderate. The posterior end ranges from sharply pointed to subrounded. The oral trough becomes shorter and shallower anteriorly with growth. Stratigraphically specimens are not abundant, but are found from basal through the middle Morrow part of the Bird Spring formation.

Material.—59 specimens.

Types.—Hypotypes Nos. 43496–43498.

Genus LIGONODINA Ulrich and Bassler in Bassler, 1925

Type species.—*Ligonodina pectinata* Ulrich and Bassler in Bassler, 1925, original designation.

Ligonodina spp.
(Pl. 8, figs. 9–12)

Remarks.—No complete specimens of *Ligonodina* spp. were found in this study. Specimens are present throughout the stratigraphic section but are most abundant in the Indian Springs formation. Four specimens are illustrated to show forms present. Specific identification was not attempted because of the fragmentary nature of the specimens.

Material.—88 specimens.

Types.—Hypotypes nos. 43534–43537.

Genus LONCHODINA Ulrich and Bassler in Bassler, 1925

Type species.—*Lonchodina typicalis* Ulrich and Bassler in Bassler ,1925, original designation.

Lonchodina spp.
(Pl. 8, figs. 2–5)

Remarks.—Specimens of *Lonchodina* spp. occur throughout the section of study but are most abundant in the Indian Springs formation. No complete individuals were found; consequently no specific identifications are attempted. Four specimens are illustrated to show forms found.

Material.—140 specimens.

Types.—Hypotypes nos. 43538–43541.

Genus MAGNILATERELLA Rexroad and Collinson, 1963

Type species.—*Magnilaterella robusta* Rexroad and Collinson, 1963, original designation.

Magnilaterella robusta Rexroad and Collinson
(Pl. 4, fig. 12)

1941*a* *Lonchodina* sp. Branson and Mehl (part), p. 171, pl. 5, fig. 10 only.

1956 *Metalonchodina?* sp. Elias, p. 124, pl. 4, fig. 3.

1957 Genus indeterminate Rexroad (part), p. 42, pl. 4, figs. 19–21 only.

1958 Genus indeterminate Rexroad, p. 26, pl. 5, figs. 1, 2.

1963 *Magnilaterella robusta* Rexroad and Collinson, pp. 14–17, pl. 2, figs. 4, 5, 9; text figs. 3*a–c*; 4*a–e*.

1964 *Magnilaterella robusta* Rexroad and Collinson. Rexroad and Furnish, p. 673, pl. 111, figs. 27–31.

Description.—See Rexroad and Collinson (1963, pp. 14–17).

Remarks.—Specimens of *Magnilaterella robusta* are rare in the section studied and were found only in the Battleship Wash formation. The lower extent of the range of this species in southern Nevada remains to be determined. Discovery of these specimens widens the geographic range of *M. robusta* which had previously been reported only from Middle and Upper Mississippian strata of the mid-continent.

Material.—3 specimens.

Type.—Hypotype no. 43454.

Genus METALONCHODINA Branson and Mehl, 1941

Type species.—*Prioniodus bidentata* Gunnell, 1931, original designation.

Metalonchodina sp.
(Pl. 8, fig. 6)

Remarks.—*Metalonchodina* sp. most closely resembles *M. tenora,* a species reported from the Desmoines series of the midcontinent (Ellison, 1941), in that it has two large denticles on the anterior limb; it differs from the latter in that it has one small denticle anterior to the two large denticles. Because two fragmentary specimens from the "marker" conglomerate of the Indian Springs formation were the only ones found, a specific designation is not given.

Material.—2 specimens.

Type.—Hypotype no. 43533.

Genus NEOPRINONIODUS Rhodes and Müller, 1956

Type species.—*Prioniodus conjunctus* Gunnell, 1931, original designation.

Neoprioniodus cf. *N. camurus* Rexroad
(Pl. 8, fig. 13)

Remarks.—All specimens of *Neoprioniodus* cf. *N. camurus* are incomplete. Specimens were found in the Indian Springs and Bird Spring formations, middle Chester through Derry. *N. camurus* has been recorded previously from Middle and Upper Mississippian rocks of the midcontinent. Additional work is needed to establish the range in southern Nevada, but it appears that *N.* cf. *N. camurus* may range higher in southern Nevada than *N. camurus* does in the midcontinent.

Material.—7 specimens.

Type.—Hypotype no. 43516.

Neoprioniodus loxus Rexroad
(Pl. 7, fig. 12)

1957 *Neoprioniodus loxus* Rexroad, p. 34, pl. 2, figs. 8, 9, 14.
1957 *Neoprioniodus tenuis* Rexroad, p. 35, pl. 2, figs. 13, 16.
1958 *Neoprioniodus loxus* Rexroad, p. 23, pl. 5, figs. 5–6.
1961 *Neoprioniodus loxus* Rexroad. Rexroad and Burton, p. 1155, pl. 140, fig. 12.
1964 *Neoprioniodus loxus* Rexroad. Rexroad and Furnish, p. 674, pl. 111, fig. 26.

Description.—See Rexroad (1957, p. 34).

Remarks.—Although of little apparent stratigraphic value, specimens of *Neoprioniodus loxus* are sparsely present throughout the section of study. Rexroad and Collinson (1961) reported a Middle and Late Mississippian range for *N. loxus* in the midcontinent. Discovery of specimens in the *Fusulinella* Zone in the Bird Spring formation extends the stratigraphic range through Derry time and the geographic range to include southern Nevada.

Material.—16 specimens.

Type.—Hypotype no. 43517.

Neoprioniodus scitulus (Branson and Mehl)
(Pl. 7, fig. 13)

1941*a* *Prioniodus scitulus* Branson and Mehl, p. 173, pl. v, figs. 5–6.
1947 *Prioniodus scitulus* Branson and Mehl. Cooper, p. 92, pl. 20, figs. 1–3.
1956 *Prioniodus scitulus* Branson and Mehl. Elias, p. 109, pl. ii, figs. 9–10.

1957 *Neoprioniodus scitulus* (Branson and Mehl). Rexroad, p. 35, pl. 2, figs. 22, 26.
1957 *Neoprioniodus striatus* Rexroad, p. 35, pl. 2, figs. 11, 12.
1957 *Neoprioniodus* sp. B. Rexroad, p. 36, pl. 2, fig. 24.
1958 *Neoprioniodus scitulus* (Branson and Mehl). Rexroad, pp. 23–24, pl. 5, figs. 10–14.
1959 *Neoprioniodus erectus* Rexroad. Elias (part), pp. 150–151, pl. 2, figs. 10–11 only.
1959 *Neoprioniodus scitulus* (Branson and Mehl). Elias, pp. 151–152, pl. 2, figs. 6–7.
1959 *Neoprioniodus cassilaris* (Branson and Mehl). Elias (part), pp. 153–154, pl. 2, figs. 20–21 only.
1960 *Neoprioniodus brevis* Clarke, pp. 13–14, pl. ii, fig. 7.
1961 *Neoprioniodus scitulus* (Branson and Mehl). Rexroad and Burton, p. 1155, pl. 140, figs. 15–17.
1961 *Neoprioniodus scitulus* (Branson and Mehl). Higgins, pl. xi, fig. 1.
1964 *Neoprioniodus scitulus* (Branson and Mehl). Rexroad and Furnish, p. 674, pl. 111, figs. 36–37.
1966 *Neoprioniodus erectus* Rexroad. Elias, p. 19, pl. 2, figs. 10, 11.
1966 *Neoprioniodus scitulus* (Branson and Mehl). Elias, p. 20, pl. 2, figs. 6, 7.
1966 *Neoprioniodus cassilaris* (Branson and Mehl). Elias, p. 23, pl. 2, figs. 17–21.
1967 *Neoprioniodus scitulus* (Branson and Mehl). Globensky, p. 443, pl. 55, figs. 22, 26.
1967 *Neoprioniodus scitulus* (Branson and Mehl). Koike (part), p. 307, pl. 4, fig. 31 only.

Description.—See Branson and Mehl (1941*a*, p. 173).

Remarks.—Rare specimens of *Neoprioniodus scitulus* are found in the Battleship Wash and Indian Springs formations. Rexroad and Collinson (1961, p. 8) discuss some of the variations among adult specimens of *N. scitulus.* These variations are apparent in the few specimens studied here and their broad interpretation of the species is accepted.

Material.—5 specimens.

Type.—Hypotype no. 43518.

Neoprioniodus singularis (Hass)

(Pl. 7, fig. 14)

1941 *Prioniodus barbatus* Branson and Mehl. Ellison and Graves (part), pl. 1, fig. 25 only.
1953 *Prioniodus singularis* Hass, p. 88, pl. 16, fig. 4.
1956 *Prioniodus roundyi* var. *dividen* Elias, p. 110, pl. ii, figs. 39–41.
1956 *Prioniodus roundyi* var. *parviden* Elias, p. 112, pl. ii, figs. 42–43.
1956 *Prioniodus* cf. *singularis* Hass. Elias, p. 112, pl. ii, fig. 45.
1957 *Prioniodina alatoidea* (Cooper). Bischoff, p. 45, pl. 5, figs. 33, 34, 36.
1958 *Neoprioniodus singularis* (Hass). Stanley, pp. 471–472, pl. 66, figs. 2–3.
1961 *Neoprioniodus singularis* (Hass). Higgins, pl. xi, fig. 5.
1962 *Neoprioniodus singularis* (Hass). Higgins, pl. 1, fig. 8.
1964 *Neoprioniodus singularis* (Hass). Rexroad and Furnish, p. 674, pl. 111, fig. 33.
1966 *Neoprioniodus singularis* (Hass). Globensky, p. 444, pl. 55, figs. 23, 24.
1967 *Neoprioniodus singularis* (Hass). Koike, p. 307, pl. 4, fig. 30.

Description.—See Hass (1953, p. 88).

Remarks.—Specimens of *Neoprioniodus singularis* are most abundant in the Battleship Wash and lower half of the Indian Springs formations. They compare closely with the original description and illustrations (Hass, 1953). Rexroad and Collinson (1961) report a Middle Mississippian into Early Pennsylvanian range for *N. singularis.* In southern Nevada the lower extent of the range was not determined and the species was not found above the Indian Springs formation.

Material.—25 specimens.

Type.—Hypotype no. 43519.

Neoprioniodus? expandofundus Webster, n. sp.
(Pl. 7, figs. 15, 16)

Diagnosis.—This species has a greatly expanded, teardrop-shaped basal cavity that is rounded anteriorly. The main cusp is discrete projecting anteriorly and then recurving posteriorly. The posterior bar bears two or three discrete denticles with the posteriormost larger than the anterior one or two.

Description.—Unit consists of main cusp, weakly developed posterior bar with two or three denticles, and greatly expanded basal cavity. Main cusp tapered, elongate, projects anteriorly then recurves posteriorly flexed inward, rounded in section at base, becoming laterally compressed distally, subrounded tip. Posterior bar short, proximal part merges smoothly with main cusp, bears two or three denticles; denticles discrete, posteriormost longest, points posteriorly, anterior one or two erect nodelike on posterior flank of main cusp. Basal cavity shallow with deep central pit, teardrop-shaped in outline, rounded anteriorly, pointed posteriorly, covers entire aboral surface, bilaterally symmetrical, central pit beneath posterior edge of main cusp, faint medial groove posterior to central pit.

Remarks.—*Neoprioniodus? expandofundus* is questionably referred to the genus because of the weakly developed posterior bar and the greatly expanded basal cavity. This species probably developed during Morrow time from a neoprioniodid similar to *N. scitulus* by suppression of the posterior bar and anticusp and expansion of the basal cavity. No described Late Paleozoic genus except *Neoprioniodus* is similar to *N.? expandofundus* and a new genus may be represented by this species.

Only three specimens of *N.? expandofundus,* one right form and two left forms, were found. One specimen from Section G was found in the lower part of the *Profusulinella* Zone; the other two specimens, one each from sections B and I, were found in the uppermost part of the *Profusulinella* Zone. Thus an early Derry range is recognized in southern Nevada for this species.

The specific name, *expandofundus,* is taken from Latin, *expando* meaning spread out and *fundus* meaning base, in reference to the expanded basal cavity.

Material.—3 specimens.

Types.—Holotype no. 43520; paratypes nos. 43521–43522.

Genus OZARKODINA Branson and Mehl, 1933

Type species.—*Ozarkodina typica* Branson and Mehl, 1933, original designation.

Ozarkodina campbelli Youngquist and Heezen
(Pl. 7, fig. 9)

1948 *Ozarkodina campbelli* Youngquist and Heezen, p. 771, pl. 118. fig. 1.

Description.—See Youngquist and Heezen (1949, p. 771).

Remarks.—*Ozarkodina campbelli* belongs to the *O. delicatula–O. compressa* group; it is distinguished by the unequal size of its denticles and by less oral arching. Variation in *O. campbelli* is most apparent in the relative height of the posterior blade, which ranges from a narrow to a moderately high blade below the denticles. The anterior blade is fairly constant and always higher than the posterior blade. There is some variation in the degree of oral-aboral arching but not sufficient to be of specific importance.

O. campbelli is found in the Bird Spring formation from middle Morrow through the *Fusulinella* Zone. Youngquist and Heezen (1949) originally reported *O. camp-*

belli from a shale bed of possible pre-Desmoines age near Knoxville, Iowa. Additional work is needed to establish the stratigraphic and geographic range of this species.

Material.—26 specimens.

Type.—Hypotype no. 43523.

Ozarkodina curvata Rexroad
(Pl. 7, fig. 10)

1958 *Ozarkodina curvata* Rexroad, p. 24, pl. 4, figs. 1–3.
1961 *Ozarkodina curvata* Rexroad. Rexroad and Burton, p. 1156, pl. 141, figs. 13–14.
1964 *Ozarkodina curvata* Rexroad. Rexroad and Furnish, p. 674, pl. 111, figs. 10–11.
1966 *Ozarkodina curvata* Rexroad. Globensky, p. 446, pl. 56, fig. 20.
1967 *Ozarkodina* spp. Koike (part), pl. 4, figs. 33, 35 only.

Description.—See Rexroad (1958, p. 24).

Remarks.—*Ozarkodina curvata* is easily identified by the pronounced oral arching and lateral bowing. Specimens were found in the Battleship Wash and Indian Springs formations. *O. curvata* appears to be a good Middle through Late Mississippian index for North America; additional work is needed, however, to establish the lower extent of its range in southern Nevada.

Material.—10 specimens.

Type.—Hypotype no. 43524.

Ozarkodina delicatula (Stauffer and Plummer)
(Pl. 7, fig. 11)

1932 *Bryantodus delicatulus* Stauffer and Plummer, p. 29, pl. 2, fig. 27.
1932 *Bryantodus nasutus* Stauffer and Plummer, p. 29, pl. 2, fig. 28.
1932 *Bryantodus sulcatus* Stauffer and Plummer, p. 30, pl. 2, figs. 11, 14, 30.
1933 *Bryantodus delicatus* Gunnell, p. 267, pl. 32, fig. 43.
1933 *Bryantodus rugosus* Gunnell, p. 268, pl. 32, fig. 44.
1933 *Bryantodus strigatus* Gunnell, p. 268, pl. 32, fig. 45.
1933 *Bryantodus strigillatus* Gunnell, p. 268, pl. 32, fig. 46.
1933 [Not] *Bryantodus delicatus* Branson and Mehl, p. 220, pl. 16, fig. 19.
1941 *Ozarkodina delicatula* (Stauffer and Plummer). Ellison, p. 120, pl. 20, figs. 40–42, 47.
1941 *Ozarkodina delicatula* (Stauffer and Plummer). Ellison and Graves, pp. 3–4, pl. 1, figs. 12–14.
1948 *Ozarkodina delicatula* (Stauffer and Plummer). Youngquist and Heezen, pp. 771–772, pl. 118, fig. 6.
1949 *Ozarkodina delicatula* (Stauffer and Plummer). Youngquist and Downs, pp. 168–169, pl. 30, figs. 1, 3, 7, 11.
1949 [Not] *Bryantodus delicatus* Branson and Mehl. Beckman, p. 161, pl. 1, fig. 7; pl. 3, fig. 3.
1952 *Ozarkodina* component Rhodes, p. 893, pl. 126, figs. 2, 3, 8–11; text fig. 2.
1957 *Ozarkodina delicatula* (Stauffer and Plummer). Bischoff, p. 39, pl. 1, figs. 25–28.
1957 *Ozarkodina delicatula* (Stauffer and Plummer). Flügel and Ziegler, p. 45, pl. 5, fig. 2.
1961 *Ozarkodina delicatula* Ellison (sic). Higgins, pl. xii, fig. 13.
1961 *Ozarkodina delicatula* (Stauffer and Plummer). Rexroad and Burton, p. 1156, pl. 141, fig. 12.
1965 *Ozarkodina delicatula* (Stauffer and Plummer). Murray and Chronic, p. 605, pl. 72, figs. 26–27.
1965 *Ozarkodina delicatula* (Stauffer and Plummer). Dunn, p. 1149, pl. 140, fig. 18.
1967 [Not] *Ozarkodina delicatula* (Stauffer and Plummer). Globensky, p. 446, pl. 56, fig. 19.

Description.—See Stauffer and Plummer (1932, p. 29); Ellison (1941, p. 120).

Remarks.—The small, numerous, subequal denticles of *Ozarkodina delicatula* easily distinguish it from *O. campbelli* and *O. curvata.* Specimens of *O. delicatula* were found throughout the section of study but are not abundant and are typically fragmentary.

Material.—49 specimens.

Type.—Hypotype no. 43525.

Ozarkodina roundyi (Hass)
(Pl. 7, fig. 8)

1926 *Ctenognathus* sp. B Roundy, p. 16, pl. 2, figs. 4–5.
1953 *Subbryantodus roundyi* Hass, p. 89, pl. 14, figs. 3–6.
1957 *Ozarkodina roundyi* (Hass). Bischoff, p. 40, pl. 1, figs. 25–28.
1957 *Ozarkodina roundyi* (Hass). Rexroad, p. 37, pl. 2, fig. 7.
1965 *Ozarkodina roundyi* (Hass). Dunn, p. 1149, pl. 140, figs. 19–20.

Description.—See Hass (1953, p. 89).

Remarks.—Specimens of *Ozarkodina roundyi* are not abundant, are usually fragmented, and occur in the Battleship Wash and Indian Springs formations. Dunn (1965) noted the gradational sequence between *O. delicatula* and *O. roundyi* when he reported both sypecies from Lee Canyon, Nevada. *O. roundyi* is known only from rocks of Chester age from North America.

Material.—13 specimens.

Type.—Hypotype no. 43526.

Genus ROUNDYA Hass, 1953

Type species.—*Roundya barnettana* Hass, 1953, original designation.

Roundya sp.
(Pl. 8, figs. 7, 8)

Remarks.—Four fragmentary specimens of *Roundya* sp., three from the Indian Springs formation and one from the upper Morrow part of the Bird Spring formation, show the expanded basal cavity typical of this genus. None is complete enough to permit unquestioned specific identification. Two specimens are illustrated to show the forms present.

Material.—4 specimens.

Types.—Hypotypes nos. 43531–43532.

Genus SPATHOGNATHODUS Branson and Mehl, 1941

Type species.—*Spathodus primus* Branson and Mehl, 1933, original designation.

Spathognathodus campbelli Rexroad
(Pl. 7, fig. 5)

1957 *Spathognathodus campbelli* Rexroad, p. 37, pl. 3, figs. 13–15.
1958 *Spathognathodus campbelli* Rexroad, p. 25, pl. 6, fig. 9.
1960 *Spathognathodus pusillus* Clarke, pp. 20–21, pl. iii, figs. 10, 11.
1961 *Spathognathodus campbelli* Rexroad. Rexroad and Burton, p. 1156, pl. 141, fig. 15.
1964 *Spathognathodus campbelli* Rexroad. Rexroad and Furnish, p. 674, pl. 111, figs. 23–24.
1966 *Spathognathodus campbelli* Rexroad. Globensky, p. 447, pl. 55, figs. 13, 20.
1967 *Spathognathodus campbelli* Rexroad. Koike, p. 310, pl. 3, figs. 26–34.

Description.—See Rexroad (1957, p. 37).

Remarks.—The occurrence of *Spathognathodus campbelli* in the Battleship Wash formation and not in the overlying Indian Springs formation indicates that this species may be restricted to Chester rocks in southern Nevada. The lower extent of its range was not determined.

Material.—188 specimens.

Type.—Hypotype no. 43507.

Spathognathodus coloradoensis Murray and Chronic
(Pl. 7, fig. 7)

1965 *Spathognathodus coloradoensis* Murray and Chronic, pp. 606–607, pl. 72, figs. 11–13.
1967 *Spathognathodus coloradoensis* Murray and Chronic. Koike, p. 310, pl. 3, figs. 23–24.

Description.—See Murray and Chronic (1965, p. 606).

Remarks.—Murray and Chronic (1965) described *Spathognathodus coloradoensis* from Desmoines strata of northwestern Colorado. Specimens from southern Nevada were found in the upper part of the *Profusulinella* Zone and throughout the *Fusulinella* Zone; the upper range limit was not determined. Specimens closely match the original description and illustrations. Discovery of the southern Nevada specimens extends the stratigraphic range to include the Derry series and the geographic range to include southern Nevada.

Material.—7 specimens.

Type.—Hypotype no. 43508.

Spathognathodus minutus (Ellison)
(Pl. 7, fig. 4)

1941 *Spathodus minutus* Ellison, p. 120, pl. 20, figs. 50–52.
1941 [?] *Spathognathodus minutus* (Ellison). Ellison and Graves, pl. 2, figs. 1, 3, 5.
1949 *Spathognathodus minutus* (Ellison). Youngquist and Downs, p. 169, pl. 30, fig. 4.
1949 *Spathognathodus minutus* (Ellison). Sturgeon and Youngquist, p. 385, pl. 74, figs. 9–11; pl. 75, fig. 19.
1960 *Spathognathodus minutus?* (Ellison). Clarke (part), p. 20, pl. iii, fig. 9 only.
1961 *Spathognathodus minutus* (Ellison). Rexroad and Burton, pp. 1156–1157, pl. 141, figs. 10–11.
1965 *Spathognathodus minutus* (Ellison). Murray and Chronic, p. 606, pl. 72, figs. 29, 30.
1965 *Spathognathodus minutus* (Ellison). Igo and Koike (part), pp. 88–89, pl. 9, figs. 16, 18 only.
1967 *Spathognathodus minutus* (Ellison). Koike, p. 311, pl. 3, figs. 39–42.

Description.—See Ellison (1941, p. 120).

Remarks.—Specimens of *Spathognathodus minutus* were not found in abundance but occur throughout the stratigraphic section of this study. Some variation was noted in the degree of aboral arching but this is thought to be a variation within the species. Illustrated fragmentary specimens from the Dimple limestone of Early Pennsylvanian age in the Marathon region of western Texas (Ellison and Graves, 1941) are questionably included in synonomy here because none of the specimens show the presence of secondary denticles along the anterior margin, and Ellison and Graves make no comment about the specimens.

Material.—45 specimens.

Type.—Hypotype no. 43509.

Spathognathodus rexroadi Webster, n. sp.
(Pl. 7, figs. 1–3)

Diagnosis.—A large species with two or three secondary denticles on the anterior margin and up to fifteen subequal denticles in the posterior blade. The posterior blade is gently bowed inward and the basal cavity is narrow and elongate.

Description.—In oral view unit narrow, elongate; anterior blade straight, posterior blade gently bowed inward; third or fourth denticle most prominent; basal cavity lips project slightly beyond side of blade posterior to main denticle.

In lateral view unit length slightly more than twice height; oral surface denticulate, composed of bluntly pointed, closely appressed denticles fused nearly to tips; main denticle distinctly wider and higher than other denticles, with two or three secondary denticles on anterior margin; denticles posterior to main denticle subequal (fifteen in mature specimens), point orally anteriorly, become increasingly reclined toward posterior tip; main basal cavity slightly anterior to midline; aboral surface sinuous, flat to orally recurved beneath main denticle and anterior end, concave posteriorly with apex of arch near posterior end of expanded basal cavity.

In aboral view basal cavity slightly less than one-third length of unit, elongate, moderately deep; situated slightly anterior of midlength of unit; both lips of cavity flared most anteriorly, inner lip less steep-walled than outer; basal cavity continues as faint medial groove on aboral surface anteriorly, as gradually narrowing moderately deep medial groove on aboral surface posteriorly.

Remarks.—*Spathognathodus rexroadi* superficially resembles *S. minutus* with the secondary denticles on the anterior margin; it differs, however, from the latter in that the posterior half of the blade is bowed inward, there are more denticles posterior to the main denticle in mature forms, the anterior tip of the aboral surface is flat to orally recurved, and the basal cavity is more elongate and less widely flared.

Specimens of *S. rexroadi* are abundant in the Battleship Wash formation and rare in the basal part of the Indian Springs formation; the lower extent of its range was not determined.

The species is named in honor of Dr. Carl B. Rexroad, whose conodont studies in the midcontinent were most helpful to me.

Material.—28 specimens.

Types.—Holotype no. 43510; paratypes nos. 43511–43514

Spathognathodus sp.
(Pl. 7, fig. 6)

Remarks.—A single specimen from the Indian Springs formation easily identified as belonging to the genus *Spathognathodus* is illustrated to show the nature of the blade denticles. The blade is formed of nine subequal denticles in a moderate arch. The apex of the arch is the fourth denticle from the anterior end. This specimen lacks the large anterior denticle and has the basal cavity situated farther toward the posterior than *S. minutus.*

Material.—1 specimen.

Type.—Hypotype no. 43515.

Genus STREPTOGNATHODUS Stauffer and Plummer, 1932

Type species.—*Streptognathodus excelsus* Stauffer and Plummer, 1932, original designation.

Remarks.—Species of *Streptognathodus* studied here suggest that the genus may be polyphyletic. Ellison (1941) stated that *Streptognathodus* and *Idiognathodus* were supposed to be derived from *Spathognathodus.* Rexroad and Burton (1961, pl. 138, figs. 7, 8) illustrated specimens transitional between *Cavusgnathus unicornis* (*C. gigantus* of this report) and *Streptognathodus unicornis;* transitional specimens between these two species were also found in southern Nevada. *S. lanceolatus* Webster, n. sp., is believed to have been derived from *S. unicornis* by nondevelopment of the prominent posterior denticle of the blade. *S. noduliferus* was probably derived from a gnathodid in Early Pennsylvanian time by coalescence of the posterior three-fourths of the main carina with the outer parapet. *S. suberectus* may have evolved from *S. noduliferus* by development of paired transverse ridges mutually extending to the median trough; alternately it may have evolved from a gnathodid by development of the transverse ridges and nondevelopment of the main carina to form a median trough. Growth series are needed to determine which of these two possibilities is the actual derivation. *S. striatus* Webster, n. sp., probably developed from *S. suberectus* by the addition of numerous finer transverse ridges on the platform.

Declinognathodus Dunn (1966, p. 1300) is considered a junior synonym of *Streptognathodus* because the blade adjoins the platform in a central position and continues onto the platform as a medial carina. The medial carina is short, either stopping abruptly or bending to merge into the outer parapet, and a medial trough continues to the posterior tip of the platform. The twenty-nine specimens Dunn had available for study do not show the range of variation for *S. noduliferus.*

Streptognathodus expansus Igo and Koike

(Pl. 6, figs. 1–5)

1964 *Streptognathodus expansus* Igo and Koike, p. 189, pl. 28, fig. 14.
1967 *Streptognathodus expansus* Igo and Koike. Koike (part), p. 312, pl. 3, figs. 6–8, 16 only.

Emended Description.—In oral view unit lanceolate-shaped, pointed posteriorly; blade narrowly compressed, continues onto platform as medial carina for one-third blade length, abruptly terminated by transverse ridge; platform linear, three-fifths unit length, widest at approximate center of length, arched inward and slopes aborally to posterior tip from blade termination; platform surface divided longitudinally by shallow trough slightly inward of medial plane, covered with transverse ridges; transverse ridges variable in number (eleven to twenty-three, number increases with growth), low, faint coalescence toward inner side, posteriormost two to four irregular throughout growth, others straight in early stages with anterior ridges convex posteriorly and posterior ridges convex anteriorly in mature stages; inner and outer parapets subequal, three to four nodes on each, separated from blade by moderate secondary troughs, slight convergence posteriorly, terminated posteriorly by same transverse ridge terminating blade; outer accessory lobe not present in immature forms, developed in mature forms widest at junction of outer parapet and transverse ridges, extends posteriorly to anteriormost one or two transverse ridges, surface covered with one to three irregular nodes; inner accessory lobe not present in immature forms, developed with maturity, delta shaped between inner parapet and anterior three to five transverse ridges, covered with two to five nodes that tend to parallel convex outer edge and enclose one to three irregularly spaced nodes.

In lateral view unit has distinct angularity in platform at termination of blade; platform wedge-shaped, narrowest at posterior tip evenly expanding to anteriormost transverse ridge, surface slopes approximately 30° aborally from line projected along oral surface of blade; inner accessory lobe separated from platform proper by shallow trough posteriorly, merges smoothly with parapet anteriorly; outer accessory lobe merges evenly with platform posteriorly and parapet anteriorly; parapets distinctly set out from blade oral surface, merge evenly with blade aborally; blade straight aborally, increasingly higher anteriorly; denticles appressed, fused nearly to tips, larger anteriorly, posterior three or four fused into a ridge; ten denticles on holotype; basal cavity lips flare aborally, separated from platform by moderately deep trough.

In aboral view basal cavity deep, smooth, elongate, flaring laterally; inner lip steep-walled, greatest flare under accessory lobe; outer lip less steep-walled, centrally flared; basal cavity continues as faint medial groove along aboral edge of blade.

Remarks.—The emended description of *Streptognathodus expansus* was prompted by the discovery of growth stages in the same sample including immature to gerontic forms. Gerontic forms show a greater development of the accessory lobes and more numerous transverse ridges. Koike (1967) considers *Idiognathodus togashii* Igo and Koike to be an immature form of *S. expansus.* Immature forms from southern Nevada show faintly the shallow medial trough and development of finer transverse ridges, which are not shown on *I. togashii.* Therefore, *I. togashii* is here considered an immature form of *Idiognathodus.*

Only left forms of *S. expansus* were found in this study. Two specimens illustrated by Koike (1967, pl. 3, figs. 7–8) appear to be right forms.

Koike reports a middle through late Morrow range for *S. expansus* from southwestern Japan. In southern Nevada *S. expansus* was found only in middle Morrow strata.

Material.—46 specimens.

Types.—Hypotypes nos. 43490–43494.

Streptognathodus lanceolatus Webster, n. sp.

(Pl. 6, figs. 14, 15)

Diagnosis.—A small form with a moderately deep medial trough and two subequal noded parapets on the platform. The blade is straight, composed of eight subequal denticles, and bends outward at the junction with the platform to merge with the outer parapet or ends at the anterior end of the medial trough. The basal cavity is moderately deep, laterally flared, and pointed anteriorly and posteriorly.

Description.—In oral view unit narrow lanceolate; blade straight, slightly less than one-half unit length, bends outward at junction with platform to merge with outer parapet or abruptly stops at anterior end of medial trough; platform narrow, straight, sharply pointed posteriorly; parapets subequal, high, steep-walled, faintly noded along oral edge, merge posteriorly, may have one or two nodes posterior to merger; trough narrow, moderately deep; basal cavity lips visible laterally.

In lateral view unit one-third as high as long, highest anteriorly, tapers posteriorly, lowest posteriorly, aboral and oral surfaces straight to slightly arched; blade composed of eight subequal, closely appressed denticles coalesced nearly to tips, merging smoothly into platform walls.

In aboral view basal cavity smooth, elongate, moderately deep, two-thirds unit length, laterally flared, deepest toward anterior end, pointed anteriorly continuing as medial groove on free blade to anterior end, sharply pointed posteriorly; cavity lips thin, flare subequal; flare on outer lip more central; greatest flare on inner lip farther toward anterior end than on outer lip.

Remarks.—*Streptognathodus lanceolatus* resembles *S. unicornis* but differs from the latter in that it lacks the large posterior denticle on the blade and the blade may bend outward to continue as the outer parapet on the platform. *S. unicornis* may be ancestral to *S. lanceolatus.*

All specimens of *S. lanceolatus* were found in a single sample of lower middle Morrow age at Arrow Canyon.

The specific name *lanceolatus* is Latin for lancelike in reference to the shape.

Material.—3 specimens.

Types.—Holotype no. 43485; paratypes nos. 43486–43487.

Streptognathodus noduliferus (Ellison and Graves) n. comb.

(Pl. 4, figs. 7, 8)

1941 *Cavusgnathus nodulifera* Ellison and Graves, pp. 4–5, pl. 3, figs. 4, 6.
1960 *Streptognathodus parallelus* Clarke, p. 29, pl. v, figs. 6, 8, 14, 15.
1962 [?] *Cavusgnathus nodulifera* Ellison and Graves. Higgins, pl. 3, fig. 27.
1964 *Streptognathodus japonicus* Igo and Koike, p. 188, pl. 28, figs. 5–13.
1966 *Declinognathodus nevadensis* Dunn, p. 1300, pl. 158, figs. 4, 8.
1967 *Idiognathoides* aff. *I. nodulifera* (Ellison and Graves). Lane, p. 938, pl. 123, figs. 9–11, 13, 16, 17.
1967 *Gnathodus nodulifera* (Ellison and Graves). Koike, p. 297, pl. 3, figs. 9–12.

Emended description.—In oral view unit slightly and regularly arched, lanceolate-shaped; blade half again longer than platform, continues as medial carina onto anterior end of platform for one to three nodes and then abruptly stops or bends outward to merge with outer parapet; platform linear, posteriorly pointed; parapets subequal to unequal, subparallel, merging at sharply pointed to subrounded posterior; inner parapet usually straight, uniformly noded; outer parapet may be covered with a line of nodes or short transverse ridges not extending into medial trough, free for anteriormost one to three nodes, remainder merged with nodes of medial carina; main trough shallow to moderately deep medial, terminating posteriorly by coalescing parapets and anteriorly by blade; secondary trough between posterior end of blade and anterior end of outer parapet short and shallower than main trough; outer and inner lips of basal cavity flare slightly to moderately, generally extend laterally beyond parapets, outer lip usually more flared than inner.

In lateral view unit wedge-shaped, thickest anteriorly; platform high-walled, gently convex orally, oral surface more convex than aboral; anterior end of inner parapet abruptly set off from blade, anterior end of outer parapet grades into blade with slight protuberance; blade straight aborally, increasingly higher anteriorly; denticles appressed, fused nearly to tips, larger anteriorly except anteriormost one or two, variable in number, ten to twelve.

In aboral view basal cavity deep, smooth, elongate, flaring laterally; on outer side flare central, on inner side strongest flaring anteriorly; cavity begins under posterior three or four denticles of blade and continues to posterior tip of specimen; faint medial groove on anterior half of aboral edge of blade.

Remarks.—Ellison and Graves (1941) named specimens from the Dimple limestone *Cavusgnathus noduliferus*. *C. noduliferus* is transferred to *Streptognathodus* because: the blade adjoins the platform axially, the medial carina gives way to a medial trough on the platform by stopping or distinctly bending to merge with the outer parapet, and the blade is not a straight line or gentle-arch continuation of the outer parapet. Illustrations labeled *S. parallelus* by Clarke (1960) closely resemble immature specimens of *S. noduliferus* and are included in the synonomy. The specimen assigned by Higgins (1962) to *C. noduliferus* is not clearly illustrated and therefore is questionably placed in synonomy here.

Dunn (1966) considered *Declinognathus nevadensis* to differ from *S. noduliferus* by having a cleanly separated single node on the anterior end of the outer parapet. This is simply one variation common on many specimens found during this study. The anterior part of the outer parapet is variously ornamented, containing one to three nodes that may or may not be semifused into a ridgelike parapet.

Streptognathodus noduliferus is one of the most abundant species found in the middle Morrow section of southern Nevada. The range apparently is lower and middle Morrow. *S. noduliferus* has been reported from the Dimple limestone of Morrow age of the Marathon region of Texas (Ellison and Graves, 1941), the middle Namurian of northwestern Spain (Higgins, 1962), and the lower part of the Millstone grit (middle Namurian) near Edinburg, Scotland (Clarke, 1960). The restricted range of this species in North America and Europe suggests that it may be an excellent guide fossil for rocks of Early Pennsylvanian age.

Material.—1,034 specimens.

Types.—Hypotypes nos. 43488–43489.

Streptognathodus unicornis Rexroad and Burton

(Pl. 4, fig. 13)

1947 *Taphrognathus varians* Branson and Mehl. Cooper, p. 92, pl. 21, figs. 14–16.
1961 *Streptognathodus unicornis* Rexroad and Burton, p. 1157, pl. 138, figs. 1–9.
1965 *Streptognathodus unicornis* Rexroad and Burton. Dunn, p. 1149, pl. 140, figs. 5, 6, 13, 14.
1966 *Adetognathus unicornis* (Rexroad and Burton). Lane, p. 930, pl. 119, figs. 16–21.

Description.—See Rexroad and Burton (1961, p. 1157).

Remarks.—Lane (1967) demonstrated that *Streptognathodus unicornis* lacks the medial carina on the anterior part of the platform which is a definitive characteristic of the genus. Furthermore, he noted that under high magnification the blade, although adjoining the platform axially, swings or bends to merge with the outer parapet and does not continue onto the platform as a medial carina as in *S. noduliferus.* He therefore transferred *S. unicornis* to *Adetognathus. Adetognathus* is not accepted here, as explained under *Cavusgnathus.* Because the blade is medial where it adjoins the platform, *S. unicornis* is left under the original generic assignment. If it is truly a cavusgnathid, an assignment that needs additional study, then the specific name is preoccupied.

Rexroad and Burton (1961) described *Streptognathodus unicornis* from the Kinkaid formation of Chester age of the Illinois Basin. They consider that *S. unicornis* was derived from *Cavusgnathus unicornis* (*C. gigantus* herein) in late Chester time and illustrate some transitional specimens. In southern Nevada *S. unicornis* ranges from at least the top of the Battleship Wash formation to the base of the Pennsylvanian, whereas *C. gigantus* was not found below the upper shales of the Indian Springs formation. Thus *S. unicornis* appears before *C. gigantus* in southern Nevada.

Material.—101 specimens.

Type.—Hypotype no. 43495.

Streptognathodus suberectus Dunn

(Pl. 4, fig. 11)

1966 *Streptognathodus suberectus* Dunn, p. 1303, pl. 157, figs. 4–6, 10.

Description.—See Dunn (1966, p. 1303).

Remarks.—Specimens of *Streptognathodus suberectus* have only one accessory lobe, but all are broken, lacking the blade and many lack the anterior part of the platform. Otherwise, these individuals compare closely with illustrations and the original description of the species. Specimens of *S. suberectus* are from the middle Morrow part of the Bird Spring formation.

Material.—24 specimens.

Type.—Hypotype no. 43484.

Genus SYNPRIONIODINA Ulrich and Bassler in Bassler, 1925

Type species.—*Synprioniodina alternata* Ulrich and Bassler in Bassler, 1925, original designation.

Synprioniodina denticamura Rexroad and Liebe
(Pl. 8, fig. 16)

1957 *Synprioniodina* n. sp. Rexroad, p. 40, pl. 4, fig. 3.
1962 *Synprioniodina denticamura* Rexroad and Liebe, p. 513, text fig. 2*a–c*.
1964 *Synprioniodina denticamura* Rexroad and Liebe. Rexroad and Furnish, p. 675, pl. 111, fig. 34.

Description.—See Rexroad and Liebe (1962, p. 513).

Remarks.—Specimens of *Synprioniodina denticamura* are not common and were found in the Battleship Wash and Indian Springs formations. Rexroad and Liebe (1962) reported a late Meramec into early Chester range for *S. denticamura.* Thus the range of this species in southern Nevada may extend slightly higher into the Chester series than that in the midcontinent; the lower extent of the range in southern Nevada is unknown.

Synprioniodina microdenta Ellison
(Pl. 8, fig. 15)

1933 *Synprioniodina* sp. Gunnell, p. 269, pl. 31, fig. 6.
1941 *Synprioniodina microdenta* Ellison, p. 119, pl. 20, figs. 43–46.
1941 *Synprioniodina microdenta* Ellison. Ellison and Graves, p. 3, pl. 1, fig. 10.
1952 *Synprioniodina* component Rhodes, p. 892, pl. 126, figs. 4, 8–11.
1964 *Synprioniodina microdenta* Ellison. Igo and Koike, pp. 184–185, pl. 27, figs. 11–17.
1965 *Synprioniodina microdenta* Ellison. Murray and Chronic, pp. 607–608, pl. 73, figs. 12–13.
1967 *Synprioniodina microdenta* Ellison. Koike, p. 313, pl. 3, figs. 43, 44.

Description.—See Ellison (1941, p. 119).

Remarks.—Two incomplete specimens, one from the upper part of the Indian Springs formation, Section B, and one from the *Fusulinella* Zone of the Bird Spring formation, Section I, are the first reported occurrences of *Synprioniodina microdenta* below rocks of Desmoines age in North America. Igo and Koike (1964) recorded *S. microdenta* from the Omi limestone of Early Pennsylvanian age in Japan. Thus the known stratigraphic range for this species is Chester into Virgil and the geographic range is North America and Japan.

Material.—2 specimens.

Type.—Hypotype no. 43528.

APPENDIX

Stratigraphic Sections

The eleven stratigraphic sections tabulated below were measured and described during the spring of 1963. Thicknesses are in feet and were obtained by direct measurement and the Brunton-eye-height method. Lithologic descriptions were completed in the field and abbreviations are listed below.

Each section was assigned a locality number from the Paleontologic Locality Register, Department of Geology, University of California, Los Angeles. Individual units from each section bear the locality number for the particular section followed by the unit number, for example, 5245–46 signifies Arrow Canyon section, unit 46.

Abbreviations Used in Lithologic Descriptions

Abbreviation	*Description*
abdt	abundant
ang	angular
approx	approximate
aren	arenaceous
arg	argillaceous
bd(s)	bed(s)
bdd	bedded
bddg	bedding
bioclas	bioclastic
bky	blocky
bl	blue
blk	black
bnd(s)	band(s)
brach(s)	brachiopod(s)
brn	brown
bry	bryozoan
c	coarse
calc	calcareous
cgl	conglomerate
chlky	chalky
choc	chocolate
chrt(y)	chert(y)
clas	clastic
cmtd	cemented
col(s)	columnal(s)
cptxln	cryptocrystalline
crin	crinoid
dcrg	decreasing
diam	diameter
discont	discontinuous
dist	distance
dk	dark
dnse	dense
dolo	dolomite
est	estimated
expos	exposure
f	fine
ferrug	ferruginous
fiss	fissile
fos	fossil(s)
fract(s)(d)	fracture(s)(d)
frag(s)	fragment(s)
fr	fair, fairly
fri	friable
gastro	gastropod
gr(d)	grain(ed)
grn	green
gry	gray
incg	increasing
irreg	irregular
lrg	large
ls	limestone
lt	light
mass	massive
matl	material
med	medium
micac	micaceous
mtrx	matrix
nod(s)	nodule(s)
nonresist	nonresistant
ool	oolitic
orng	orange
o.s.	oil stained
p	poor
pbbl(s)(y)	pebble(s)(y)
pelecy(s)	pelecypod(s)
per	permeability
petro	petroliferous
plnr	planar
pls	plates
plty	platy
pnk(ish)	pink(ish)
por	porosity
produc(s)	productid(s)
prpl	purple
pt	part

ABBREVIATIONS USED IN LITHOLOGIC DESCRIPTIONS—Continued

Abbreviation	*Description*	*Abbreviation*	*Description*
ptg(s)	parting(s)	spltry	splintery
qtz	quartz	srtd	sorted
qtzite	quartzite	srtg	sorting
qtzitic	quartzitic	ss	sandstone
rd(d)	round(ed)	stnd	stained
resist	resistant	struct(s)	structure(s)
s/	some	thck	thick
sb	sub	thn	thin
scat	scattered	tr	trace
sd(y)	sand(y)	undulty	undulatory
sft	soft	v	very
sh	shale	w/	with
shly	shaly	wea	weathers
silicif	silicified	wh(ish)	white(ish)
sl	slightly	xbdd(g)	cross-bedded(ing)
slt(y)	silt(y)	xl(n)	crystal(line)
sml	small	yel	yellow
sol	solitary		

SECTION A. KANE SPRINGS WASH EAST

Location.—Lat. 37° 6′ 20″ N, Long. 114° 43′ 30″ W, to Lat. 37° 6′ 53″ N, Long. 114° 44′ 45″ W, Lincoln County, Nevada.

Comments.—Section measured in fault block of Paleozoic sediments bounded by Tertiary volcanics; started in Chainman shale and ended where volcanics in fault contact with lower Derry part of Bird Spring formation (fig. 17). Section lacks *Fusulinella-Chaetetes* Zone because of faulting. Minor contortion in Chainman shale and Indian Springs formation and several small faults of 2′ to 5′ displacement in Bird Spring formation. Indian Springs formation less than 50 percent exposed and Bird Spring approximately 80 percent exposed. UCLA locality 5242.

Unit	*Thickness*	*Lithologic description*
		Bird Spring formation
60	12.6	Ls, lt to med gry, wea lt gry, f to med bioclas, v f xln mtrx, thn to med bdd, p expos, calcite filled fracts, gry chrt nod in upper pt, contorted in pt.
59	13.3	Ls, lt to dk gry, wea lt bl-gry to med gry, f to med bioclas, v f xln mtrx, thn to med bdd, fr resist and expos, clacite filled fracts, gry to brn chrt nods.
58	4.7	Cover.
57	5.2	Ls, lt gry to prpl-gry, wea same, med bioclas, v f xln mtrx, thn bdd, gry chrt nods, p resist, fr expos.
56	23.9	Ls, lt to med gry, wea lt gry, v f xln to f bioclas, cptxln mtrx, thn to med bdd, sl aren at top, gry nodular chrt bnds, fr resist and expos.
55	9.8	Cover.
54	22.2	Ls, lt gry, wea same, f bioclas, v f xln to cptxln mtrx, med to thck bdd, sl aren, upper 2.8′ v aren (f ang qtz), p resist and expos, scat gry chrt nods becoming more abdt upward.
53	5.4	Ls, lt gry, wea bl-gry, f bioclas, v f xln mtrx, thn to med bdd, fr resist and expos, scat sml 1″ to 4″ chrt nods.
52	12.7	Sdy ls, lt gry, wea same to gry-tan, f bioclas, sd f ang qtz, med bdd, v f xln mtrx, p resist, lower 5′ cover.

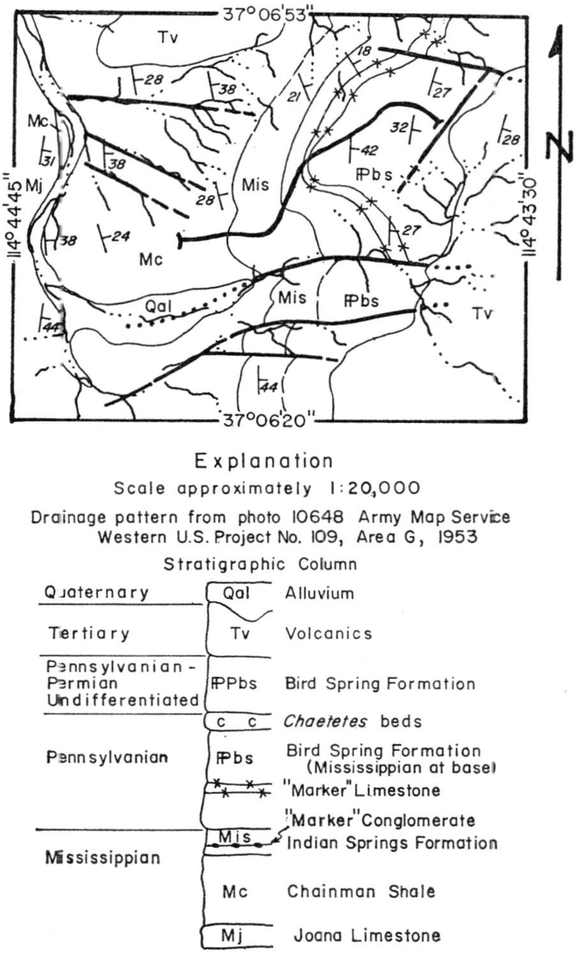

Fig. 17. Geologic map showing line of traverse for Section A, Kane Springs Wash East, Lincoln County, Nevada. See figure 21 for explanation of symbols.

Unit	*Thickness*	*Lithologic description*
51	10.2	Ls, lt gry, wea same, f to med bioclas, v f xln mtrx, gry to tan chrt nods, med bdd, fr resist and expos.
50	7.6	Cover, pnk wea ls in float.
49	7.4	Ls, lt gry, wea same, f bioclas, v aren (v f ang well srtd qtz), v f xln mtrx, med bdd, lrg chrt nods in upper pt showing slump structures?, fr resist, slope holder.
48	24.8	Ls, lower pt as lower pt of underlying unit, upper pt sdy ls, grading upward into calc cmtd ss, f ang qtz grs, well srtd, gry to cream chrt nods and discont bnds.
47	12.7	Ls, lt to med gry, wea lt gry, f to med bioclas, v f xln mtrx, med bdd, upper 3′ gry to tan chrt nods, fr resist and expos, slope holder.
46	39.6	Ls, lt to med gry, wea lt gry to gry-brn, f bioclas, v f xln mtrx, gry chrt nods and discont nodular bnds, thn bdd in lower pt, med bdd in upper pt, upper pt fr resist, cliff former.

Unit	*Thickness*	*Lithologic description*
45	6.6	Ls, med gry, wea lt gry-bl, f to med bioclas, med bdd, bky wea, fr resist and expos.
44	12.2	Ls, as below w/ *Osagia?*, nonresist, lower half p expos.
43	31.8	Ls, lt gry-tan, wea yel-orng to gry-yel, v aren and chrty, f bioclas, thn bdd, upper 12′ fr resist and cliff former, lower pt nonresist fr expos, lower pt almost entirely chrt in s/ places.
42	14.8	Ls, med gry, wea lt gry-bl, f to med bioclas, v f xln mtrx, thn to med bdd, dk gry to brn chrt nods, platy wea in pt.
41	20.9	Ls, lt gry, wea yel-tan to gry-yel, v f xln, aren (v f ang? qtz), thn bdd in thck units, abdt gry chrt nods and nodular bnds, fr resist, ledge holder, base less resist.
40	16.1	Ls, lt to med gry, wea lt gry, f to c bioclas, v f xln mtrx, lower pt appears laminated on wea surface, thn to med bdd, chrt nods in upper 1′, center p expos.
39	16.6	Ls, lt to med gry, wea lt gry, f to med bioclas, v f xln mtrx, thck to med bdd, fr resist and expos.
38	21.8	Ls, lt to med gry, wea lt gry, f to med bioclas, aren (f ang qtz), v f xln mtrx, lower 1.5′ arg, ledge holder, xbdd in middle, abdt dk gry-brn chrt nods and bnds, thn to med bdd.
37	4.4	Ls, med gry, wea lt bl-gry, f to med bioclas, v f xln mtrx, thck bdd, lt to dk gry-brn chrt nods, resist, caps underlying unit.
36	107.5	Ls, lt to med gry, wea lt gry w/ brn stripes, looks brn from distance, big cliff former, v f to med bioclas, aren (f ang sd), cptxln to v f xln mtrx, v abdt ferrug brn wea chrt nods and nodular bnds, thn bdd but bdg not distinct from any distance and wea as a mass unit, s/ silicif of fos, fos not abdt, crin cols, *Spirifer?* 57′ above base, *Straparolus?* 75′ above base, and sol corals 78′ above base.
35	21.0	Ls, med gry wea bl-gry, f to med bioclas, v f xln mtrx, thck bdd, few dk gry chrt nods at top, pit and cusp wea, abdt calcite filled fracts, cliff former.
34	45.7	Ls, lt to dk gry, wea tan to lt gry-yel at base and darkens to lt gry at top, v f to med bioclas, cptxln to v f xln mtrx, aren w/ dcrg sd content upward, scat gry to brn chrt nods, thn to med bdd, slope holder.
33	51.4	Ls, med to dk gry, wea lt gry to bl-gry, v f to med bioclas, cptxln to v f xln mtrx, thn to thck bdd, scat sml (¼″ to 2″) gry to brn chrt nods, lower 10′ cliff holder, remainder slope former, 4 phaceloid coral biostromes at 2′, 10′, 15′, and 28′ above base.
32	48.2	Alternating chrty ls and nonchrty ls, unit fr resist slope holder w/ upper and basal cliff formers, abdt calcite-filled fracts and sml faults: ls, med gry, wea lt gry-bl, f to med bioclas, v f xln mtrx, med to thck bdd, silicif *Derbyia?* 4′ above base; chrty ls, lt to dk gry, wea lt gry, v f to f bioclas, cptxln to v f xln mtrx, sml gry to brn chrt nods and discont bnds, thn to med bdd, minor contortion.
31	4.1	Cover.
30	49.1	Ls, med gry, wea lt bl-gry, f to med bioclas, v f xln mtrx, abdt dk gry-brn chrt nods in discont bnds in pt, scat silicif brach frags, med to mass bdd, resist cliff former.
29	21.6	Ls, lt to dk gry, wea lt gry, v f to f bioclas, sl arg, cptxln to v f mtrx, abdt fos, strong petroleum odor on fresh break, thn to med bdd, considerable gry-brn chrt nods w/ ferrug coating.
28	6.2	Ls, lt gry, wea same, c bioclas, v f xln mtrx, dk gry-brn chrt nods in upper half, considerable silicif of fos frags, med to thck bdd, fr resist and expos, few producs.

Unit	*Thickness*	*Lithologic description*
27	76.5	Alternating cover and ls, approx 60 percent expos; ls, lt to dk gry, wea lt gry to gry-tan, v f to med bioclas, v f xln mtrx, sl arg in pt, scat to abdt dk gry to brn chrt nods, nods in discont bnds in pt and wea ferrug-brn, s/ fos may be float from overlying slope former.
26	47.8	Ls, lt to med gry, wea lt gry w/ tan to pnk cast, f to c bioclas, v f xln mtrx, thn to thck bdd, upper half sl aren, dk gry to brn to blk chrt nods in discont bnds, upper half of thck bdd pt lack chrt, resist, cliff former sd grs v f to f ang well srtd. *Schizophoria* at top of cliffy lower half.
25	36.0	Ls, lt to med gry, wea lt gry to gry-bl, f to med bioclas, v f xln mtrx, thn to med bdd, fr resist and expos, slope holder, dk gry to blk to brn chrt nods in discont bnds.
24	85.1	Ls, lt to med gry, wea lt gry to gry-bl, f to med bioclas, v f xln to cptxln mtrx, thn to thck bdd, abdt dk gry to brn to blk chrt nods in discont bnds, sl aren in lower 12′, resist cliff holder, silicif sol corals and produc.
23	28.4	Ls, lt to med gry, wea gry-bl, f to med bioclas, v f xln mtrx, med to thck bdd, minor chrt, cliff former, resist, silicif spirifid, 2′ thck coral biostrome 8.1′ above base.
22	6.3	Ss, lt tan, wea salmon-pnk, f grd, ang to sb-ang, sb-qtzitic, p expos, fr resist, upper 1.9′ v calc and grades into overlying ls.
21	14.9	Cover, except for basal and upper foot: upper foot, sdy sh, grn to yel, plty, nonresist; lower foot, sh, grn to yel, plty, sft, nonresist.
20	37.4	Ls, as in underlying unit, w/ less chrt and thcker bdd, upper 13′ thner bdd.
19	13.9	Ls, lt to med gry, wea lt gry, f to med bioclas, v f xln mtrx, few dk brn wea discont chrty bnds, med to mass bdd, resist cliff former, first good cliff in section, silicif brachs.
18	8.6	Ls, lt to med gry, wea same w/ pnkish tint in s/ bds, thn to med bdd, f to c bioclas, v f xln mtrx, arg ptgs, fr resist and expos.
17	26.9	Cover.
16	14.7	Ls, lt to med gry-tan, f to c bioclas, v f to f xln mtrx, thn to thck bdd, fr resist and expos, arg ptgs.
	1,108.6	Total thickness of measured part of Bird Spring formation; conformable contact with Indian Springs formation.
		Indian Springs formation
15	2.0	Ls, reddish-brn, wea same, c bioclas, f xln mtrx, ferrug mtrx in pt, med bdd, fr resist, p expos.
14	14.2	Cover.
13	9.7	Ls, gry, wea gry-brn to rust-brn, med to c bioclas, f xln mtrx, thn to med bdd, arg ptgs, fr resist, upper half p expos.
12	21.0	Cover.
11	18.8	Ss, gry to rose-prpl to tan, wea reddish-brn, v f to f grd, ang to sb-rdd, fr srtg, sb-qtzitic in pt, thn to med bdd (0.5′ to 2′), xbdd (sml scale festoons), upper half p expos, fr resist.
10	67.9	Cover.
9	4.6	Ls, w/ cgl on upper 0.6′, lt gry, wea same, med to c bioclas, v f xln mtrx, mass, cgl wea prpl-red, pbbls up to 2″ diam, ferrug cmt in pt, thn bdd, conodonts; unit ledge former.
8	100.5	Cover, float suggest gry to blk sh.
	238.7	Total thickness of Indian Springs formation; conformable contact with Chainman scale.

Unit	*Thickness*	*Lithologic description*
		Chainman shale
7	4.5	Ss, dk gry, wea gry-brn, v f grd, ang?, well srtd, thn bdd, sb-qtzitic, fr resist and expos.
6	25.4	Cover.
5	4.6	Ls, cgl, gry-brn, wea rust-brn, pbbl up to 1.2″, thck bdd, calc mtrx, fr resist and expos, may be a lens, not expos along strike.
4	26.5	Sh, dk gry, wea gry-grn, plty, fiss, nonresist, upper 8′ cover.
3	25.5	Ss, lt gry, to brn, wea ferrug-orng, interbdd w/ gry-brn, f grd, ang, well srtd, thn to med bdd, xbdd (sml, 1″ to 6″ long, ⅛″ to 1″ deep festoons), ripple marks (oscillation troughs orient N 25° W), sb-qtzitic in pt, ferrug cmt in pt, ledge former, fr expos.
2	69.6	Cover mostly, few p expos are sh, dk gry to blk, wea gry-grn, plty, nonresist, *Cravenocerous* in lower 6′ to 8′.
1	6.4	Ss, gry to prpl, wea prpl-gry to orng, med bdd, v f to f grd, ang, sb-qtzitic mostly, calc cmt in pt, fr resist and expos.
	162.5	Total thickness of measured part of Chainman shale.

SECTION B. KANE SPRINGS WASH WEST

Location.—Lat. 37° 2′ 53″ N, Long. 114° 52′ 10″ W, to Lat. 37° 3′ 18″ N, Long. 114° 52′ 20″ W, Lincoln County, Nevada.

Comments.—Section measured in two segments on large fault block of Upper Paleozoics surrounded by Tertiary volcanics (fig. 18). The lower 657.3′ measured across southern part of block and upper 590.8′ measured across northern end. Exposures generally good, but Indian Springs formation faulted, contorted, and not as well exposed as Bird Spring formation. Numerous small faults with 2.0′ to 10.0′ displacement present in Bird Spring formation. UCLA locality 5243.

Unit	*Thickness*	*Lithologic description*
		Bird Spring formation
76	5.6	Ls, med gry, wea same, med bioclas, v f xln mtrx, thn to med bdd, resist, cliff holder, scat *Chaetetes* heads up to 1′ diam.
75	15.8	Ls, lt to med gry, wea lt gry, f to med bioclas, v f xln mtrx, large (up to 2′ diam) dk brn chrt nods, aren in lower pt, med to thick bdd, resist.
74	19.8	Sdy ls grading upward into ss, lt gry to tan, wea rust-orng to gry-tan, sd grs f ang qtz well srtd, med bdd in mass unit, p resist and expos.
73	10.5	Ls, lt gry, wea same, f to med bioclas, v f xln mtrx, med bdd, resist, fr expos, common *Chaetetes* heads up to 1.6′ in diam.
72	22.6	Ls, lt gry, wea same, f bioclas, cptxln to v f xln mtrx, silty, abdt dk gry to brn chrt nods w/ ferrug gossans, thn bdd in mass unit, p resist, fr expos.
71	3.3	Ls, med gry, wea lt gry, f to med bioclas, v f xln mtrx, dk gry chrt nods which wea w/ ferrug gossan, thn to med bdd, resist.
70	20.2	Ls, lt gry, wea same, basal pt wea orng-tan, f to med bioclas, v f xln mtrx, aren in lower pt, cover in pt at top, p resist and expos, thn to med bdd.
69	3.4	Ls, lt gry, wea same, f to med bioclas, v f xln mtrx, med bdd, fr resist and expos, *Chaetetes* heads up to 1′ diam, rare.
68	35.2	Ls, lt gry, wea same, f to med bioclas, silty, chrt nods and discont bnds, thn bdd in mass units, fr resist and expos, slope holder, *Fusulinella* and *Straparolus*? at 16′ to 21′ above base of unit.

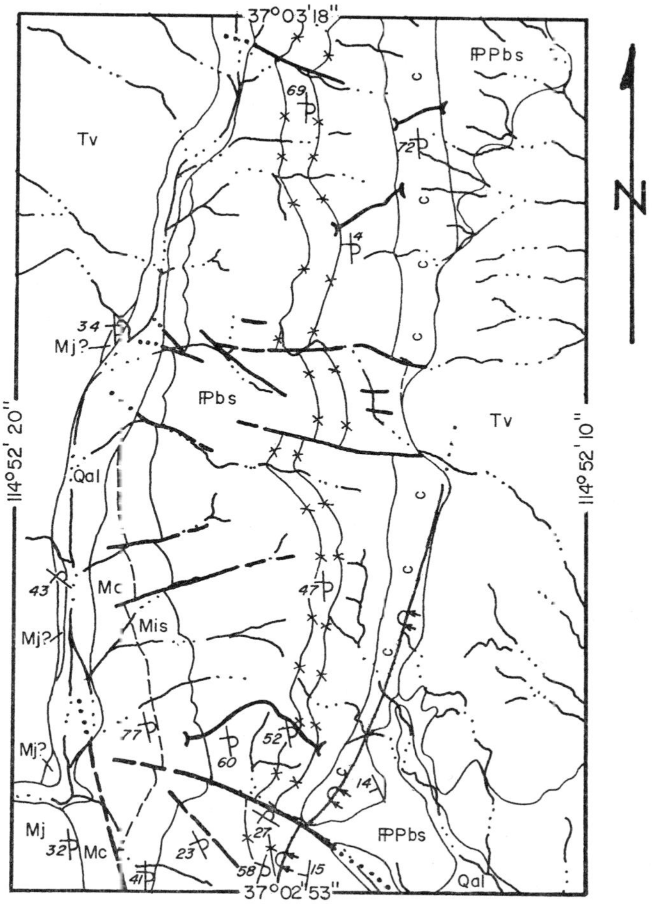

Fig. 18. Geologic map showing line of traverse for Section B, Kane Springs Wash West, Lincoln County, Nevada. See figure 17 for explanation of map units and figure 21 for explanation of symbols.

Unit	*Thickness*	*Lithologic description*
67	21.9	Ls, med gry, wea bl-gry, f to med bioclas, v f xln mtrx, dk gry to brn chrt nods and discont bnds, thck to mass bdd, resist cliff holder, *Chaetetes* heads up to 1.2′ diam in middle 4′ of unit.
66	20.9	Ls, lt gry, wea same, f to med bioclas, scat dk brn wea chrt nods, v f xln mtrx, thn to med bdd, p resist and expos.
65	32.4	Ls, lt gry, wea same, f bioclas, v f xln to cptxln mtrx, thn to med bdd, p resist, 50 percent expos.
64	9.5	Ls, lt to med gry, wea bl-gry, f to med bioclas, v f xln mtrx, few chrt nods up to 0.7′ in diam, med to thck bdd, good resist and expos, *Chaetetes* heads up to 1.3′ in upper 2′.
63	9.3	Ls, lt gry, wea same, f bioclas, cptxln to v f xln mtrx, med bdd, appears laminated on wea surface, fr resist and expos.
62	19.2	Ls, lt to med gry, wea lt gry, upper pt v f bioclas, lower pt med bioclas, cptxln to v f xln mtrx, few scat chrt nods in upper pt, lower 3.5′ laminated, thn to med bdd, p resist and expos.

Unit	Thickness	Lithologic description
61	13.5	Ls, med gry, wea bl in upper pt and dk ferrug-brn in lower 6′, f to med bioclas, v f xln mtrx, thn bdd w/ abdt chrt nods and bnds in lower 6′, med bdd w/ little chrt in upper pt, cliff holder.
60	34.6	Ls, lt to med gry, wea lt gry, f bioclas, v f xln mtrx, scat chrt nods, thn bdd in mass unit, cover in lower pt (10 percent), p resist and expos.
59	22.6	Ls, med to dk gry, wea lt gry to prpl-gry, f bioclas, arg, v f xln to cptxln mtrx, dk gry to blk chrt nods, thn bdd, wea plty to spltry to bky, p resist, fr expos.
58	50.0	Interbdd thn bdd chrty ls and med to thck bdd less chrty ls, unit resist, fr expos; med gry, wea lt bl-gry, f to med bioclas, v f xln mtrx, dk gry to brn chrt nods and discont bnds which wea w/ a ferrug gossan.
57	10.1	Ls, lt gry, wea lt gry to rust-orng, v f xln, silty, chrt nods and bnds, thn bdd, p resist and expos.
56	29.0	Ls, med gry, wea bl-gry, f to med bioclas, v f xln mtrx, little to abdt chrt nods and discont bnds, thn to med bdd, fr resist and expos.
55	12.8	Ls, lt gry, wea lt gry to rust-orng, v f bioclas, v aren (v f ang? qtz), cptxln mtrx, chrt nods and silicif nods, both wea rust-orng, thn bdd, middle pt of unit fr resist and well expos, upper and lower pts p expos.
54	36.1	Ls, lt gry, wea same, v f to f bioclas, cptxln to v f mtrx, alternating thn bdd chrty ls and med bdd sl chrty ls, cover in pt (10 percent), slope holder.
53	3.3	Cover.
52	11.2	Ss, grades upward into ls, tan to lt gry, v f to f grd, ang, well srtd, calc mtrx, wea lt gry to dk ferrug-brn, thn to med bdd, fr resist and expos.
51	22.7	Ls, lt gry, wea lt gry to gry-tan, f bioclas, aren (f ang qtz), few silicif chrt nods and bnds, upper pt less sdy, thn to med bdd, cover in pt, p resist and expos.
50	15.8	Ls, lt gry, wea lt gry, f bioclas, aren (f sbrdd qtz), chrt and silicif nods, v f xln mtrx, s/ sml xbdd, thn to med bdd, fr resist and expos, cliff holder.
49	17.5	Cover top and bottom, middle ls, lt gry, wea same, f bioclas, cptxln mtrx, thn bdd, dk gry chrt bnds and nods, p resist and expos.
48	23.3	Ls, lt to med gry, f bioclas, aren (f ang qtz), v f xln mtrx, chrt nods and discont bnds, thn to thck bdd, fr resist and expos.
47	10.6	Ls, lt gry, wea same, f bioclas, v f xln mtrx, aren as below, med to thck bdd, v resist, good expos.
46	9.8	Ls, lt gry, wea gry to gry-tan, f bioclas, v aren (f ang qtz) chrt bnds and silicif sdy bnds, thn bdd in med units, fr resist and expos.
45	7.8	Ls, med gry, wea lt bl-gry, f bioclas, cptxln to v f xln mtrx, scat gry to brn chrt nods, med bdd, fr resist and expos.
44	4.5	Ls, med gry, wea lt gry, v f to f bioclas, cptxln to v f xln mtrx, v chrty (gry to dk brn), wea plty, thn bdd, p resist and expos.
43	6.0	Cover.
42	91.7	Ls, lt to med gry, wea lt gry, f bioclas, v f xln mtrx, aren (v f ang qtz), v abdt gry to brn chrt nods and discont bnds, dcrg chrt in upper 12′, thn bdd in a mass unit, cliff former.
41	28.1	Ls, lt gry, wea same, f bioclas, v f xln mtrx, abdt chrt nods and bnds, s/ thn med gry wea bnds, thn bdd, fr resist, slope holder.
40	21.1	Ls, med gry, wea lt gry, f bioclas, v f xln mtrx, abdt chrt nods

Unit	*Thickness*	*Lithologic description*
		and buds, thn to thck bdd, 3 bds of phaceloid corals, 7′, 10′, and 16′ above base of unit, respectively, *Syringopora* at top, s/ silicif of fos.
39	20.7	Ls, med gry, wea lt gry-bl, f to med bioclas, v f xln mtrx, thn to thck bdd, scat dk gry to brn chrt nods w/ ferrug gossans, fr resist and expos, cliff holder.
38	6.5	Cover.
37	43.0	Ls, lt to dk gry, wea lt to med bl-gry, f to med bioclas, v f xln mtrx, thn to med bdd, dk gry to brn chrt nods and discont bnds w/ ferrug gossan, cgl 28′ above base (sml chrt pbbl), sl aren (f ang qtz) in lower pt, good resist and expos, s/ silicif of brachs, *Chonetes*? at top.
36	19.5	Ls, lt to med gry w/ prpl tint, wea lt yel to tan to pnkish-gry, v f xln, cptxln mtrx, arg, slty, wea plty to spltry, thn bdd, p resist and expos, abdt fos and s/ silicif of fos.
35	10.9	Ls, med gry, wea lt gry-bl, f to c bioclas, v f xln mtrx, gry to brn chrt nods which wea w/ ferrug gossan, thn to med bdd, fr resist and expos, s/ silicif of fos.
34	20.6	Cover except for 4′ in middle of unit; ls, med gry, wea lt gry, f bioclas, v f xln to cptxln mtrx, thn to med bdd, fr resist and expos, few ptly silicif producs, silicif ramose bry.
33	17.1	Ls, lt to med gry, wea bl-gry, f to med bioclas, v f xln mtrx, cgl at base, sl aren in lower 4′, thn to med bdd, fr resist and expos, gry to brn chrt nods (½″ to 2″ diam) scat through unit not v abdt, s/ silicif of fos.
32	5.6	Ls, lt gry, wea prpl-gry to yel-gry, f to med bioclas, incg sd (f ang well srtd) upward, v f xln mtrx, med bdd, fr resist and expos.
31	8.6	Cover.
30	33.9	Ls, lt to med gry, wea lt gry, f bioclas, abdt chrt nods and discont bnds, thn to med bdd, fr resist and expos, slope holder.
29	13.6	Ls, lt gry, wea same, f to med bioclas, v f xln mtrx, dcrg sd upward, chrt bnds in upper 2.4′, resist cliff former, silicif brachs in pt.
28	44.8	Ls, lt to med gry, wea lt bl-gry, f to c bioclas, v f xln mtrx, incg gry to brn chrt nods, thn to med bdd, thck bds more resist ledge formers, cover in pt in upper pt, intensely fractd w/ calcite fillings and offsets, s/ silicif of fos.
27	42.5	Ls, lt to med gry, wea lt gry-bl, f to med bioclas, v f xln mtrx, gry to brn chrt nods and discont bnds, thn to med bdd, intensely fractd w/ calcite filling of fracts, numerous sml faults w/ 0.1′ to 1′ displacement, thck bds more resist and form ledges.
26	22.1	Ls, lt to med gry, wea gry-bl, f to c bioclas, dcrg sd upward, scat chrt nods in upper half, thn to med bdd, intensely fractd w/ calcite filling in pt, fr resist and expos, phaceloid corals 14′ above base of unit, *Michelinia*? just above corals.
25	11.9	Ss, tan to lt gry, wea salmon-tan, v f to f grd, ang to sb-ang, fr well srtd, calc cmt incg upward, sb-qtzitic in lower 7′, thn to thck bdd, xbdd (sml festoons), fr resist and expos.
24	10.0	Ls, lt to med gry, wea lt gry w/ reddish cast at base, f to med bioclas, v f xln mtrx, thn to med bdd, fr resist, covered intervals.
23	40.3	Ls, lt to med gry, wea med gry, f to c bioclas, v f xln mtrx, dk gry to brn chrt nods, few chrt bnds in lower pt, thn to thck bdd, resist, fr expos, abdt fos, s/ silicif of fos.
	1,103.3	Total thickness of measured part of Bird Spring formation; conformable contact with Indian Springs formation.

Unit	*Thickness*	*Lithologic description*
		Indian Springs formation
22	5.3	Ls, lt to med gry, wea med gry at top, grades down into gry-tan, f to c bioclas, v f xln mtrx, thn to med bdd, s/ silicif of fos, chrt nods and bnds in upper half, good resist and expos.
21	10.5	Cover, w/ exception of thn ss, lt gry to red, wea same, v f grd, ang?, p resist and expos, in middle of unit.
20	12.0	Ls, lt gry wea med gry to tan, f to c bioclas, v f xln mtrx, thn bdd, poor resist, cover in pt, bellerophontid gastro in lower pt.
19	13.5	Cover.
18	4.8	Ls, lt gry to reddish-gry, wea reddish to yel-gry, med to c bioclas, v f xln mtrx, w/ s/ ferrug matl, few ferrug nods up to ⅜″, fr resist and expos, thn to med bdd.
17	8.7	Cover.
16	1.3	Ls, reddish-gry, wea reddish to yel-gry, f bioclas, v f xln mtrx, ferrug nods up to ⅛″ diam, fr resist and expos, med bdd.
15	6.4	Cover.
14	1.6	Ls, lt yel to tan, wea rust-yel to red, med to c bioclas, ferrug and v f xln mtrx, thn bdd, p expos.
13	5.5	Cover.
12	3.2	Ls, lt gry-yel, wea same, med to c bioclas, v f xln mtrx, thn bdd, fr resist and expos, calcite filled fracts.
11	3.8	Cover.
10	0.8	Ss, prpl-red, wea hematite red-blk, v f grd, sb-qtzitic, sml festoon xbdd (1″ to 2″ long, ⅛″ to ¼″ deep), thn bdd, p expos.
9	6.4	Ls, med gry, wea lt gry to reddish-pnk, f to med bioclas, v f xln mtrx, w/ s/ ferrug mtrx, thn bdd, p expos, covered interbds.
8	6.3	Ls, med prpl-gry to yel-gry, lower 1.1′ wea rusty gry-yel, upper 3.1′ wea med gry, f to c bioclas, v f xln mtrx, thn to med bdd, fr resist and expos.
7	7.9	Cover.
6	3.3	Ls, reddish-gry, wea rust-red to gry-tan, med to c bioclas, ferrug and v f xln mtrx, thn bdd, dk red ferrug gossan at base, fr resist and expos.
5	20.7	Cover.
4	2.0	Ls, gry-brn to rust-brn, med to c bioclas, f xln and ferrug mtrx, wea rust-brn w/ ferrug gossan at base, med bdd, fr resist and expos.
3	17.4	Cover.
2	1.9	Ls, yel-brn, wea same, med to c bioclas, v f xln and ferrug mtrx, med bdd, fr resist and expos.
1	1.5	Ls, orng-brn to dk red-brn, wea rusty red-brn, f to med bioclas, ferrug and v f xln mtrx, thn bdd, p resist and expos.
	144.8	Total measured Indian Springs formation; base not exposed.

SECTION C. MEADOW VALLEY RANGE

Location.—S ¼, sec. 19, N ¼, sec. 30, T. 13 S., R. 65 E., Clark County, Nevada.

Comments.—Section poorly exposed and numerous offsets along bedding planes required to avoid small faults and folds (fig. 19). Section measured for thickness and facies information. UCLA locality 5244.

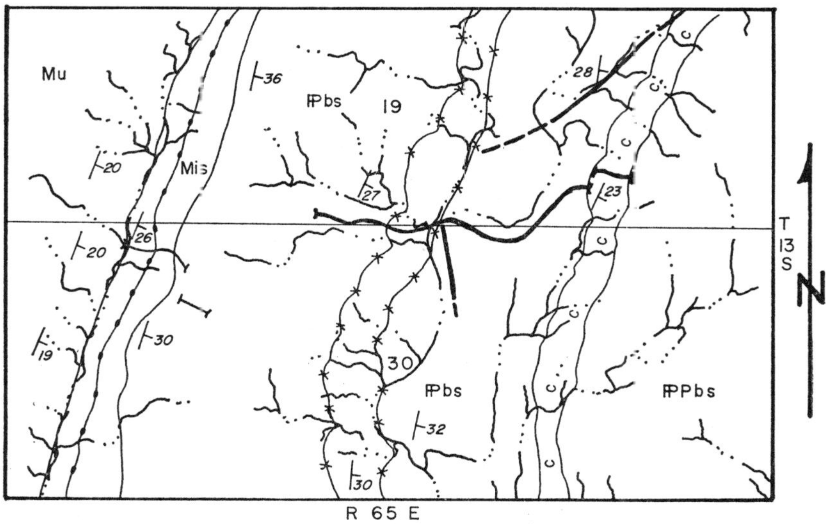

Fig. 19. Geologic map showing line of traverse for Section C, Meadow Valley, Clark County, Nevada. See figure 21 for explanation of symbols.

Unit	*Thickness*	*Lithologic description*
		Bird Spring formation
37	6.6	Ls, lt gry, wea bl-gry, v f to f bioclas, cptxln mtrx, med bdd, s/ *Chaetetes* heads in lower 2.4′, phaceloid corals in upper 4.2′, fr resist and expos.
36	111.3	Cover 90 percent; few thn p resist lt gry f bioclas ls expos.
35	10.0	Ls, as in unit 33; *Chaetetes* heads up to 1.5′ fr abdt.
34	60.5	Cover, 98 percent; two thn wh wea lt gry ls bds 20′ above base of unit, v f bioclas, cptxln mtrx.
33	12.5	Ls, lt gry, wea bl-gry, v f to f bioclas, cptxln mtrx, thn to med bdd, resist ledge former, *Chaetetes* heads up to 1′ diam, and *Multithecapora*? in middle of unit, *Fusulinella* immediately above *Chaetetes*.
32	30.3	Cover 90 percent; few thn ls bds as underlying unit.
31	86.9	Mostly cover, 80 percent; expos ls, lt gry, wea same, f to med bioclas, few dk brn chrt nods, thn bdd, *Profusulinella*? in upper 4′.
30	4.8	Ls, lt gry, wea same, f to med bioclas, thck bdd, stylolites, dk brn chrt in upper 1′, s/ reworking.
29	6.1	Ls, v aren, lt gry-tan, wea rust-yel, v f bioclas, cptxln mtrx, sd grs v f sb-ang well srtd, dk brn chrt nods and discont bnds, thn bdd, fr resist and expos.
28	51.8	Ls as in underlying unit with alternating cover intervals, 60 percent expos.
27	51.4	Ls, lt gry to med gry, wea lt gry to tan, v f to f bioclas, cptxln mtrx, s/ dk brn chrt nods, thck to mass bdd, lower 6′ wea whish, upper 13′ sml scale xbdd w/ worm borings 14′ from top of unit, resist cliff former.
26	202.5	Alternating cover and expos ls, 30 percent expos; ls, lt to med gry, wea lt gry to tan, v f to f bioclas, cptxln mtrx, s/ dk brn chrt nods, most chrty bds v aren (v f to f sb-ang qtz), thn to med bdd, float in cover intervals suggests sl arg and aren ls.

Unit	*Thickness*	*Lithologic description*
25	36.4	Cover.
24	98.8	Ls, lt gry, wea rust-brn to tan, grades into lmy ss in places, v aren, v f to f sb-ang well srtd qtz, abdt dk brn chrt nods, resist big cliff unit v thn bdd but bdg not distinct from any distance, wea as a mass unit, few fos frags.
23	71.5	Ls, lt gry to med bl-gry, wea lt gry, f to c bioclas, cptxln mtrx, scat dk brn chrt nods and s/ discont bnds, thn to thck bdd, fr resist and expos, few silicif corals and brachs.
22	23.9	Ls, lt to dk gry, wea tan to lt gry-yel at base and darkens to lt gry at top, v f to med bioclas, cptxln to v f xln mtrx, aren w/ dcrg sd content upward, scat gry to brn chrt nods, med bdd, slope holder.
21	31.8	Ls, med to dk gry, wea lt gry to bl-gry, v f to med bioclas, cptxln to v f xln mtrx, thn to thck bdd, scat sml (¼″ to 2″) gry to brn chrt nods, fr resist and expos.
20	7.6	Cover.
19	114.0	Ls, lt to med gry, wea lt bl-gry, v f to f to med bioclas, sl aren (v f to f sb-ang qtz), cptxln to v f xln mtrx, dk brn chrt nods and discont bnds, thn to med bdd, fr resist, 90 percent expos.
18	5.7	Ls, lt gry, wea med bl-gry, v f bioclas, cptxln mtrx, dk brn chrt nods, abdt silicif *Orthotetes* particularly at the base of the unit w/ barnacle? borings, thck bdd, resist.
17	10.6	Cover.
16	48.8	Ls, lt to med gry, wea same, f to med bioclas, cptxln to v f xln mtrx, thn to thck bdd, s/ dk brn to blk chrt nods and discont bnds, alternating resist and nonresist bds, silicif producs, *Osagia* in upper 2′, s/ pit and cusp wea.
15	51.0	Alternating ls as in underlying unit and cover; less than 25 percent expos.
14	79.0	Ls, lt to med gry, wea same, v f to c bioclas, cptxln to v f xln mtrx, dk brn to blk chrt nods and discont bnds, thn to thck bdd, 90 percent expos, resist to nonresist, cliff former in general. *Osagia* in upper 4′.
13	8.0	Ss, wh to lt gry, wea tan, f grd, well srtd, sb-ang to sb-rdd, calc mtrx in pt, siliceous in pt, sml scale xbdd, fr resist, med bdd.
12	16.2	Ls, lt gry, wea med gry, f to med bioclas, cptxln mtrx, sl aren, thn bdd, p resist and expos.
11	24.2	Ls, lt to med gry, wea med gry, f to med bioclas, cptxln mtrx, dk brn chrt nods and discont bnds up to 4″ thck, med to thck bdd, resist, cliff former.
10	22.6	Cover.
	1,284.8	Total thickness of measured part of Bird Spring formation conformable with Indian Springs formation.
		Indian Springs formation
9	34.1	Ls, gry to tan to yel, wea same, grades upward into more gry ls, f to c bioclas, thn to med bdd, p resist and expos.
8	45.9	Cover, probably sh.
7	3.9	Ls, tan to gry, wea yel-red, f to med bioclas, v ferrug mtrx, sb-plty, nonresist, p expos.
6	21.6	Cover, probably sh.
5	4.0	Ls, dk gry-red, wea ferrug-red, lower pt gry, med bioclas ls, v f xln and ferrug mtrx, upper pt cgl, chrt and ls clasts up to 3″ diam, ferrug mtrx, thn to med bdd, resist, fr expos.
4	20.3	Cover, probably sh.

Unit	Thickness	Lithologic description
3	29.8	Sh, varicolor, red to wh to gry to yel, wea maroon-red, plty to bky, crumbly, sft, fiss in pt, p resist, interbdd thn resist ls, tan to yel to red, v ferrug, f to med bioclas, sl aren in pt, v f ang qtz, p expos.
2	41.2	Cover, probably sh.
	200.8	Total thickness of Indian Springs formation; conformable with Battleship Wash formation.
		Battleship Wash formation
1	Not measured	Ls, lt gry, wea bl-gry, med to c bioclas, cptxln mtrx, med to mass bdd, resist, cliff former.

SECTION D. ARROW CANYON

Location.—S ½, sec. 12, T. 14 S., R. 64 E., Clark County, Nevada.

Comments.—Section measured in Arrow Canyon along north and south walls depending on occurrence of best outcrops (fig. 20). Samples taken from both walls and traced laterally into line of section. Some contortion and small scale faulting in Indian Springs formation; two small faults and one related fold in Bird Spring formation. All structural features observed only in north wall of canyon. Indian Springs outcrops fair to good; Bird Spring outcrops excellent. UCLA locality 5245.

Unit	Thickness	Lithologic description
		Bird Spring formation
108	2.4	Ls, dk gry, wea lt gry, bioclas, upper 0.6′ arg, v f xln mtrx, med bdd, petro odor on fresh break.
107	29.4	Interbdd ls and shly ls: ls, lt to med gry, wea lt gry, bioclas, v f xln mtrx, dolo and chrt nods, bddg 1.0′ to 2.2′, bddg surface sb-plnr to undulty, resist; shly ls, lt gry, wea yel-gry, bky to plty, nonresist.
106	9.9	Shly ls, lt gry, wea same, f xln, v arg, bky to plty, blk chrt nods, fr resist.
105	20.3	Interbdd ls and shly ls: ls, lt to med gry, wea lt gry, bioclas, v f xln mtrx, dolo nods, resist bddg 0.6′ to 2.0′, surface undulty, upper ls aren; shly ls, lt gry, wea yel-gry, plty, dolo nods, nonresist, p expos.
104	4.9	Ls, med gry, wea lt gry, bioclas, v f xln mtrx, bddg 0.4′ to 1.6′, surface undulty, resist, dolo nods in upper 1.0′.
103	7.6	Ls, lt gry, wea yel-gry, bioclas, v f xln mtrx, discont dolo nods 1″ to 3″ in diam, arg, thck bdd w/ shly ptgs, nonresist, abdt fos in upper 1.0′.
102	18.1	Ls, med gry, wea lt gry to gry-brn, bioclas, v f xln mtrx, aren and xbdd in lower 5.5′, bddg 0.4′ to 4.0′, surface sb-plnr to undulty, scat ferrug stnd dolo nods, s/ pit and cusp wea, resist.
101	5.0	Shly ls, lt gry, wea yel-gry, spltry, bioclas, v arg, abdt fos frags, lower half cover.
100	3.6	Ls, lt gry, wea yel-gry, bioclas, cptxln mtrx, thck bdd, wea rdd, nonresist, p expos, *Chaetetes* heads up to 2.0′ in diam.
99	14.7	Ls, lt to med gry, wea lt gry, bioclas, v f to f xln mtrx, thn to thck bdd, 0.3′ to 4.1′, resist, cliff former, scat gry dolo nods in discont bnds near middle and top.
98	6.0	Ls, med gry, wea lt gry, bioclas, f xln mtrx, bddg 0.2′ to 1.0′, surface even, resist, v abdt ramose and fenestrate bryozoan, produc with spines attached.

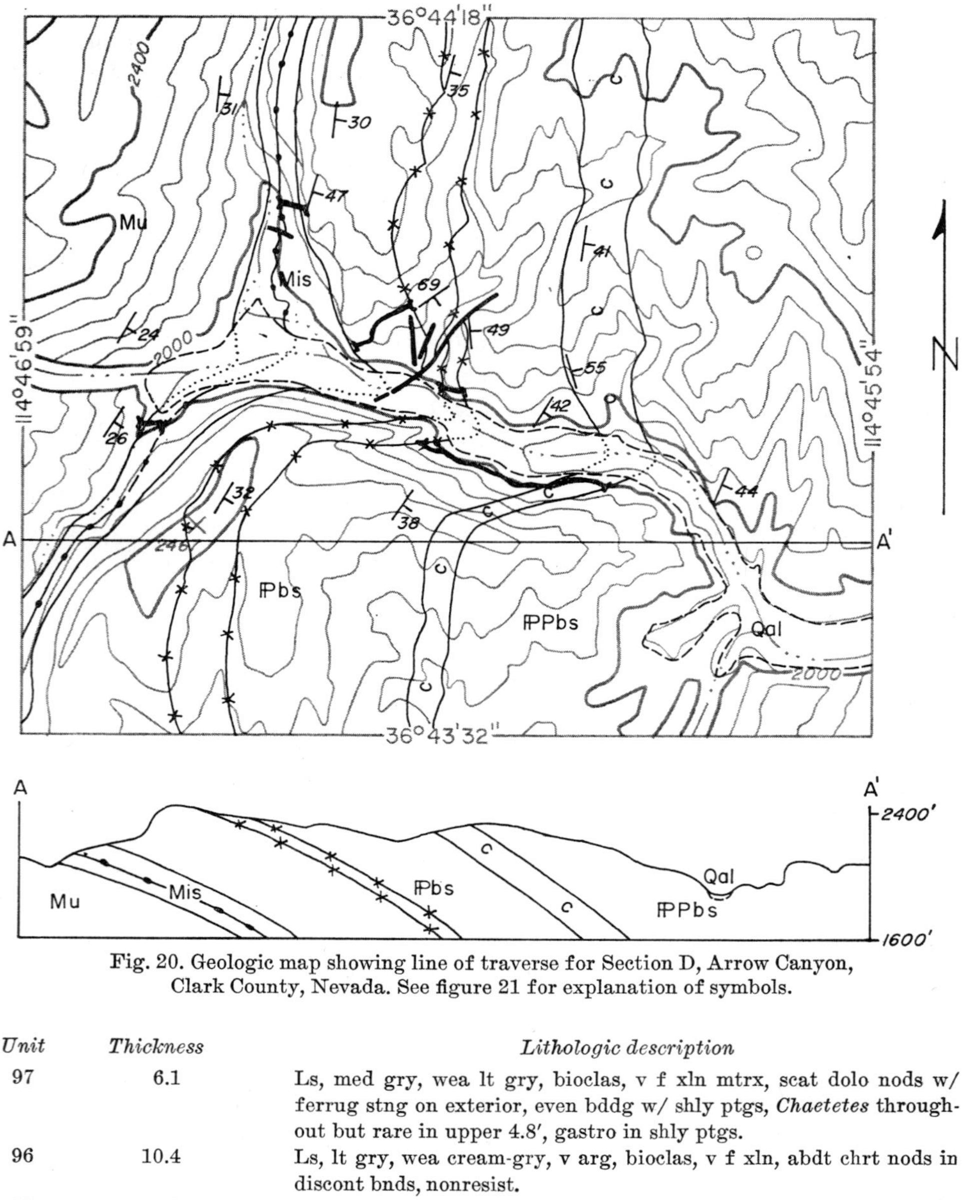

Fig. 20. Geologic map showing line of traverse for Section D, Arrow Canyon, Clark County, Nevada. See figure 21 for explanation of symbols.

Unit	*Thickness*	*Lithologic description*
97	6.1	Ls, med gry, wea lt gry, bioclas, v f xln mtrx, scat dolo nods w/ ferrug stng on exterior, even bddg w/ shly ptgs, *Chaetetes* throughout but rare in upper 4.8′, gastro in shly ptgs.
96	10.4	Ls, lt gry, wea cream-gry, v arg, bioclas, v f xln, abdt chrt nods in discont bnds, nonresist.
95	7.2	Ls, lt to med gry, wea lt gry, bioclas, v f xln mtrx, lower 3.8′ thck bdd w/few *Chaetetes* and assoc *Syringopora?* s/spiriferids, produc, ramose and fenestrate bryozoans, *Fusulinella* in upper 0.3 of lower pt; upper 3.4′ w/ arg ptgs, chrt nods, abdt *Fusulinella*.
94	5.7	Shly ls, lt gry-grn, wea lt gry, bky, nonresist, lower half covered, fos frags.
93	1.9	Ls, lt gry, wea same, bioclas, v f xln mtrx, even bddg, blk chrt nods in discont bnds.
92	4.3	Cover.

Unit	*Thickness*	*Lithologic description*
91	14.4	Ls, lt to med gry, wea lt gry, bioclas, f xln mtrx, thck bdd, cliff former, 1.2′ thck shly bd 2.7′ above base w/ spiriferids, produc; upper 4.0′ w/ produc, spiriferids, bryozoans.
90	19.3	Limy sh, grading upward into ls, lt gry, wea rust-tan, abdt chrt and dolo nods, thn bdd, wea in a mass unit, lower ⅔ nonresist, upper ⅓ ledge former, few fos in lower 3.0′.
89	27.5	Ls, lt to med gry, wea lt gry, bioclas, v f xln to cptxln mtrx, aren in lower 4.0′, scat chrt and dolo nods w/ ferrug stng on exterior, nods up to 1.0′ in diam in uppermost bed, lower ½ thn bdd w/ arg interbds, spiriferids, produc, *Composita* in arg bds.
88	22.9	Ls, med gry, wea lt gry to bl-gry, bioclas, sl aren and arg in lower 6.4′, v f xln to cptxln mtrx, s/ pit and cusp wea, bddg thn to thck, 0.1′ to 5.5′, surface sb-plnr to undulty.
87	37.3	Ls, lt to med gry, wea lt gry to whish, bioclas, arg, v f xln mtrx, bddg thn in mass units to thn bdd, 0.2′ to 2.0′, shly ptgs, p expos, abdt chrt and dolo nods, dolo most common, ferrug gossans on nods, sol corals v abdt in lower 7.1′, *Profusulinella* in next 3.8′ w/ abdt crin pls and rare cups, few produc and spiriferids in remainder of unit.
86	19.2	Ss, lt gry to tan, wea rust-orng-brn, v f to f grd, ang, fr srtg, calc cmt, p cmtd in lower 6.0′, 1.5′ transition of sdy ls at base, discont bddg, 0.5″ to 3.1′ sml scale xbddg, cliff former.
85	11.8	Ls, lt gry, wea gry to brn, aren, thn bdd in mass unit w/ abdt nod discont chrt bnds, arg and p expos at base, resist, cliff former.
84	12.9	Ls, lt gry, wea gry-bl, bioclas, cptxln mtrx, pit and cusp wea, uniform bddg, 0.3′ to 2.0′, sb-plnr surface, resist, cliff former, silicif brachs in upper 3.0′.
83	12.5	Ls, lt gry, wea same, bioclas, arg w/ chrt bnds at base and top, f xln mtrx, thn bdd, middle pt med bdd, p expos, nonresist, large 1.5″ diam ramose bryozoans in lower pt.
82	8.6	Ls, med gry, wea lt gry, bioclas, v f xln mtrx, aren in lower 2.4′, scat rdd chrt nods up to 1.0′ in diam, bddg 0.8′ to 2.4′ surface sb-plnr, resist, cliff former.
81	8.0	Ls, lt gry, wea buff, bioclas, arg, chlky to v f xln mtrx, v abdt blk discont nod chrt bnds, p resist, shly ptgs, thn bdd.
80	30.2	Ls, lt to med gry, wea lt gry-tan, bioclas, cptxln to v f xln mtrx, aren (v f grd ang qtz), scat dk brn to blk chrt nods and discont bnds, bddg undulty, 1.0′ to 4.0′, cliff former.
79	3.8	Ls, lt gry, wea gry cream, bioclas, v f xln mtrx, arg in pt, bddg 0.3′ to 1.0′, p expos.
78	9.8	Ls, lt to med gry, wea lt gry, bioclas, v f xln mtrx, scat dk brn chrt nods and bnds, med bdd.
77	11.3	Ls, lt gry, wea cream to tan, v aren (v f grd ang qtz), thck bdd, fr resist.
76	7.3	Ls, lt to med gry, wea gry-brn, bioclas, aren (v f ang qtz), abdt chrt nods and bnds, thck bdd.
75	6.2	Ls, lt to med gry, wea lt gry, bioclas, v f xln mtrx, scat chrt nods, thck bdd, resist.
74	6.3	Ls, lt gry, wea gry-brn, v aren and chrty, thn discont, bddg in mass unit, resist.
73	18.4	Ls, lt gry, wea same, bioclas, chlky to f xln mtrx, slty, arg, s/ shly ls, blk discont nod chrt bnds, thn to med bdd, 0.2′ to 1.8′, bddg sb-plnr, nonresist, p expos.

Unit	Thickness	Lithologic description
72	5.4	Ls, lt to med gry, wea lt gry to brn, bioclas, cptxln to v f xln mtrx, dk brn to blk discont nod chrt bnds, resist.
71	7.5	Ls, lt to med gry, wea lt gry, bioclas, v f to f xln, cptxln mtrx, dnse, mass.
70	7.0	Ls, lt gry, wea dk brn to rust-brn, bioclas, v f xln mtrx, arg, thn discont bddg in mass unit, s/ silicif fos.
69	13.1	Ls, lt to med gry, wea lt gry, bioclas, cptxln to v f mtrx, few chrt bnds, bddg 1.0′ to 3.0′, shly ptgs.
68	7.9	Ls, lt gry, wea dk brn to rust-brn, bioclas, v f xln mtrx, arg, thn discont bddg in mass unit, s/ silicif of fos.
67	8.2	Ls, lt gry, wea cream-gry, v f xln, cptxln mtrx, arg, slty, dk brn to blk chrt bnds 0.5″ to 4.0″, qtz filled vugs, no chrt in upper 2.0′, thn bdd, nonresist.
66	24.0	Ls, lt gry, wea same, bioclas, aren and arg in pt (sd grs f, ang, well srtd qtz), s/ xbddg, chrt bnds dk brn to blk nod discont, bddg fr regular, 0.2′ to 3.0′, sb-plnr, 1.2′ sh near base, lower 7.0′ wea yel.
65	24.2	Chrty ss, lt gry, wea rust-orng, v f to f grd, ang, well srtd, calc mtrx, thn bdd in mass units, upper 2.7′ wea dk brn, lower 6.8′ v lmy, grades into sdy ls, wea gry, dk brn to blk nod discont chrt bnds, 0.5″ to 5.0″, resist.
64	4.0	Ss, lt gry, wea lt rust-orng, f grd, ang, well srtd, med scale xbddg, 2.0′ to 3.0′ long, 0.2′ to 0.6′ high, thn bdd, calc mtrx.
63	10.8	Ls, lt gry to bl-gry, wea yel in lower 5.0′ and lt gry in upper pt, bioclas, v f to med xln, incr xln size upward, cptxln mtrx, shly interbds, dk brn to blk discont nod chrt bnds in upper half, nonresist.
62	13.9	Ls, lt to med gry, wea lt gry grading upward into dk rust-brn, bioclas, f to med xln, v f xln mtrx, aren, f ang qtz, dk brn to blk discont nod chrt bnds, med scale xbddg in aren pt, 1.0′ long, 2.0″ deep, thn to med bddg, 0.6′ to 3.0′, undulty surface, resist.
61	16.9	Ls, lt to med gry, wea lt gry, bioclas, v f xln to cptxln mtrx, upper 7.2′ aren (f grd ang qtz), dk brn to blk discont nod chrt bnds, arg, bddg papery to 1.4′, sb-plnr surface, scat fos.
60	3.2	Calc sh, lt gry, wea same, nonresist, lowermost 0.3′ grades into bioclas ls.
59	92.3	Ls, lt to med gry, wea gry-brn to rust-brn, bioclas, v f to f xln, cptxln mtrx, v chrty, nod layered dk brn to blk, discont bnds, aren (f grd ang qtz), thn discont bddg in mass unit, v resist, cliff former, spiriferids, produc, echinoid spines, crin frags; 6.8′ arg ls 20′ below top, wea lt yel-orng, nonresist.
58	23.8	Ls, lt to dk gry, wea bl-gry, bioclas, f to med xln, minor c xln, cptxln to v f xln mtrx, scat dk brn chrt nods, aren in middle, thck bdd, cliff former, silicif brach and corals.
57	33.8	Ls, lt to dk gry to blk, wea gry-brn, bioclas, cptxln to v f xln mtrx, v chrty in thn 1.0″ to 3.0″ discont nod bnds, bddg thn discont in mass units, 1 shly bd 7.0′ above base, another 7.0′ below top.
56	13.0	Ls, lt to med gry, wea lt gry, bioclas, med to c xln, v f xln mtrx, thck discont bddg, calc filled fracts, s/ pit and cusp wea, resist.
55	17.7	Ls, med gry, wea lt gry, bioclas, cptxln to v f xln mtrx, aren below middle w/ sml slump structs in aren pt, med to thck bdd, 0.7′ to 2.0′, bddg s/ what discont, stylolitic contacts in pt, resist, cliff former, 3 layers of phaceloid corals, base, 6.0′ below top, and 3.0′ below top, respectively.

Unit	*Thickness*	*Lithologic description*
54	16.3	Ls, med gry, wea lt olive-gry-brn, cptxln to v f xln mtrx, abdt dk gry to blk chrt nods in discont bnds, bddg discont, thck, resist, cliff former.
53	11.7	Ls, lt to med gry, wea lt gry, bioclas, f to med xln, v f xln mtrx, arg in upper 5.4′, irreg brn chrt nod in lower 6.3′, stylolites, thn to thck bdd, resist, s/ silicif of fos.
52	9.4	Ls, lt cream-gry, wea same, bioclas, v f xln, arg, grades into lmy sh, nonresist.
51	24.0	Ls, med gry, wea olive-brn-gry, bioclas, v f xln mtrx, abdt dk brn to blk discont chrt bnds, bddg p developed, thn to med bdd in mass unit, radiating xls, resist, scat silicif brach.
50	0.4	Cgl, lt gry, wea salt and pepper gry and blk, calc mtrx, calc pbbl up to 1.0″ diam, rdd, p srtg, f to med qtz grs, irreg bddg, 0.2′ to 0.6′.
49	11.0	Ls, lt gry, wea med bl-gry, bioclas, f to c xln, cptxln to v f xln mtrx, aren in pt, wea tan where aren, dk brn to blk discont nod chrt bnds, p resist.
48	26.7	Ls, lt to dk gry, wea bl-gry, bioclas, v arg at base, c clas upward, incg dk brn chrt nods in discont bnds upward, resist, cliff former, scat silicif fos frags.
47	12.2	Ls, lt to dk gry, wea gry-tan, bioclas, med bioclas grading upward into v f xln and arg ls, dk brn discont chrty bnds, shly ptgs, silicif brach in lower ½, produc and *Composita* in shly ptgs.
46	18.8	Interbdd ls and calc sh: ls, lt gry, wea same, bioclas, cptxln to v f xln mtrx, arg, thn to med bdd, 0.6′ to 1.2′, surface sb-plnr, abdt fos as in sh; sh, lt gry-grn, wea yel-gry, bky to plty, v calc, p resist and expos, abdt fos.
45	14.6	Ls, med to dk gry, bioclas, cptxln mtrx, arg, shly ptgs, dk brn to blk discont nod chrt bnds in upper bddg surfaces, bddg 0.3′ to 2.4′, sb-plnr surface, s/ silicif of fos.
44	1.9	Sh, lt gry, wea yel-gry, plty to bky, nonresist.
43	4.6	Ls, lt gry, wea same, bioclas, cptxln mtrx, arg, ferrug stng along worm? tubs, nonresist, p expos, 0.6′ ss bd at top (v f to f grd, ang, well srtd, calc mtrx), ss looks like bioclas ls on fresh break, wea orng-brn.
42	7.9	Ls, lt to med gry, wea med gry, bioclas, f to med xln, cptxln mtrx, resist, cliff former, silicif fos frags; dk brn calc ss 3.0′ from base, 0.2′ to 0.6′ thck, shows slumping, wea in relief; dk brn discont 0.3′ chrty bnd 1.0′ below top.
41	5.6	Interbdd ls and shly ls: ls, med gry, wea same, bioclas, f to med xln, cptxln mtrx, thn bdd, chrt nods and wavy bddg in upper ls; shly ls, lt gry, wea yel-gry, sft, nonresist, p expos.
40	2.7	Ls, lt gry, wea same, bioclas, f to med xln, cptxln mtrx, scat brn chrt nods in lower pt, bddg irreg and discont, abdt silicif brachs.
39	25.1	Ls, lt to med gry, wea lt gry, bioclas, f to med xln, arg and cptxln mtrx, shly ptgs, abdt discont nod dk brn chrt bnds, chrt decreases upward, bddg 0.3′ to 3.0′, irreg but persist, fr resist, scat fos most abdt in ptgs.
38	3.8	Ls, med gry, wea lt gry, bioclas, f xln mtrx, thck bdd, resist, *Osagia*? (or oolites?).
37	6.6	Ls, med gry, wea lt gry, bioclas, dk brn chrt nods w/ ferrug gossan, shly interbds in upper 4.0′, fr resist, s/ silicif of fos, especially in lower bd.
36	13.8	Ls, dk gry, wea med gry, bioclas, v f xln at base grading upward

Unit	*Thickness*	*Lithologic description*
		into med xln, cptxln mtrx, discont nod chrt bnds in lower 4.0′, strong fetid odor on fresh break, bddg 0.2′ to 3.0′, discont in pt, sb-plnr surface, cliff former.
35	12.3	Ls, lt gry, wea gry-tan, bioclas, med to c xln, v f xln mtrx, discont nod dk brn chrt bnds, bddg 0.3′ to 3.0′, sb-plnr surface, cliff former, silicif fos.
34	1.9	Sh, lt gry-grn to prpl-bl, wea lt yel-gry, v calc, nonresist, p expos.
33	26.8	Ls, lt to med gry, wea lt gry, bioclas, f to c xln, cptxln mtrx, discont nod dk brn to blk chrt bnds more abdt upward, 0.2′ to 0.5′, bddg 0.5′ to 4.0′, thn bdd upward, sb-plnr surface, silicif of fos in pt.
32	15.1	Ls, lt gry, wea same, bioclas, med to c xln, v f xln mtrx, 3.4′ tan wea unit 2.1′ below top, scat dk brn chrt nods 1″ to 3″ in diam in upper 2.1′, xbdd, stylolitic contacts, petro odor on fresh break, s/ silicif of fos.
31	5.1	Lmy ss, lt gry, wea lt rust-brn, f grd, ang, well srtd, calc mtrx, xbdd (3.0′ to 4.0′ long, 0.3′ to 0.7′ high), looks like bioclas ls on fresh break.
30	4.1	Sh, lt yel-gry, wea yel, bky, nonresist, 4.0″ aren stringers near base and top.
29	1.0	Ls cgl, choc-brn, ls pbbl up to 1.5″ diam, rdd clasts, even bddg.
28	5.8	Ls, lt gry, wea same, bioclas, calcite filled fracts, thn bdd (0.1′ to 1.0′), discont bddg, s/ silicif of fos.
27	2.6	Shly ls, lt gry, wea yel-gry, bioclas p expos, abdt brachs.
26	23.0	Ls, lt to med gry, wea same, bioclas, brn to blk discont chrt nod bnds, petro odor on fresh break, bddg 0.1′ to 1.3′, sb-plnr surface, shly ptgs, cliff former, s/ silicif of fos.
25	0.4	Calc sh, lt gry, plty, sft.
24	14.8	Ls, lt to med gry, wea lt gry w/ s/ rusty wea in lower 5.0′, bioclas, f to c xln, cptxln mtrx, few discont nod brn to blk chrt bnds, shly interval 1.5′, slightly above middle of unit, thn bdd, 0.05′ to 1.0′, discont, s/ sml to med scale xbddg, s/ silicif of fos, abdt fos.
	1,102.0	Total thickness of measured part of Bird Spring formation; conformable contact with Indian Springs formation.
		Indian Springs formation
23	3.8	Ls, buff-yel, wea same, bioclas, thn bdd, nonresist.
22	2.7	Ls, lt gry-brn, wea rust-brn, bioclas, cptxln mtrx, med to thck bdd, abdt spiriferids on upper surface.
21	4.8	Sh, lt grn to red, wea yel to red, plty to bky, nonresist, thn ss bd near top, lt gry-brn, wea rust-brn, v f grd, sb-ang, well srtd, calc mtrx, sml scale xbddg, resist.
20	7.0	Ls, lt gry, wea rust-yel, bioclas, f to c xln, v f xln mtrx, shly near base, med bdd, sb-plnr surface, resist.
19	4.8	Ss, lt gry-brn, wea rust-brn, v f grd, sbang, well srtd, calc mtrx, sml scale xbddg, v f ferrug spotting, mass, resist.
18	4.9	Pbbly ss, lt brn, wea same, v f grd, thn bdd, 0.5″ to 4.0″, pbbl mostly rdd, fos frags.
17	4.5	Sh, lt yel-gry, wea yel-tan, plty to bky, nonresist, grades into underlying sdy sh.
16	3.2	Sdy sh, lt yel-gry, wea same, sft, punky, nonresist, v f grd ang qtz.
15	9.0	Coquina cgl, lt rust-yel, wea same, med to pbbl size clasts, sb to well rdd, fri, mostly fos frags, med bdd, wavy and xbdd, s/ channeling at base, up to 1.0′.
14	13.9	Ss, lt gry-tan, wea tan to rust-brn, v f to f grd, sb-ang, well srtd,

Unit	*Thickness*	*Lithologic description*
		calc mtrx, sml scale xbddg, current ripple marks, E-W along trough, bddg laminae to 1.0′ sb-plnr surface, p resist.
13	5.8	Ls, lt yel-gry, wea same, bioclas, shly ptgs, bddg 0.3′ to 1.6′, s/ ferrug stng, abdt produc.
12	30.8	Mostly cover, less than ⅓ expos, sh, lt gry-yel to maroon, wea same, plty to bky, sft, nonresist.
11	3.7	Ls, lt yel-brn, wea rust-brn, bioclas, cptxln mtrx, ferrug, thn bddg in upper pt, thck bdd below, resist.
10	22.4	Sh, varicolored, med gry to yel to prpl to red, wea lt gry-yel, plty to spltry, brittle, nonresist, cover in pt.
9	3.9	Ls, gry to tan, wea gry to rust-brn, bioclas, thck bdd, calcite filled fracts, scat produc, upper 0.8′ cgl, dk ferrug red, wea dk rust-brn to blk, mass, flakey on top, silicious and ls nod, up to 4.0″ in diam, mostly pbbl size to 2.0″.
8	17.6	Sh, varicolored, bl-gry to red, wea red, plty to bky, nonresist, cover in pt.
7	3.7	Ls, yel-brn, wea rust-red, bioclas, dk ferrug-red upper surface, calc filled fracts, thck bdd.
6	26.5	Sh, med gry, wea maroon, plty to bky, nonresist, cover in pt, petro distilled by boiling.
5	12.5	Sh, med gry, wea maroon, bky to plty, nonresist, p expos.
4	1.3	Ls, lt yel-gry, wea rust-yel-red, bioclas, f to c xln, cptxln mtrx, ferrug, even bddg, thck bdd, dk brn ferrug gossan on top surface, worm ? trails.
3	5.2	Sh, dk gry to blk, wea maroon, plty to bky, sft, brittle, sl calc, nonresist.
2	28.0	Cover.
	220.0	Total thickness of Indian Springs formation; conformable contact with Battleship Wash formation.
		Battleship Wash formation
1	Not measured	Ls, gry-brn, wea lt bl-gry, f to c bioclas, cptxln mtrx, v irreg upper surface, med to mass bdd, abdt fos.

SECTION E. BATTLESHIP WASH

Location.—Center S ½, sec. 26 to Center SW ¼, sec. 25, T. 14 S., R. 64 E., Clark County, Nevada.

Comments.—Section measured in east trending tributary canyon to Battleship Wash (fig. 21), Indian Springs formation poorly exposed and contorted. Bird Spring formation well exposed. Few small faults on both sides of canyon which required several offsets in line of section. Samples collected on both walls of canyon, depending on occurrence of best outcrop, and traced laterally into line of section. UCLA locality 5246.

Unit	*Thickness*	*Lithologic description*
		Bird Spring formation
111	4.6	Ls, lt to med gry, wea lt gry, f to med bioclas, lower ½ appears laminated on wea surface, sl aren thn to med bdd, fr resist and expos.
110	6.6	Cover.
109	10.6	Ls, lt gry, wea same, f to c bioclas, abdt brn chrt nods, thn to med bdd, fr resist and expos.

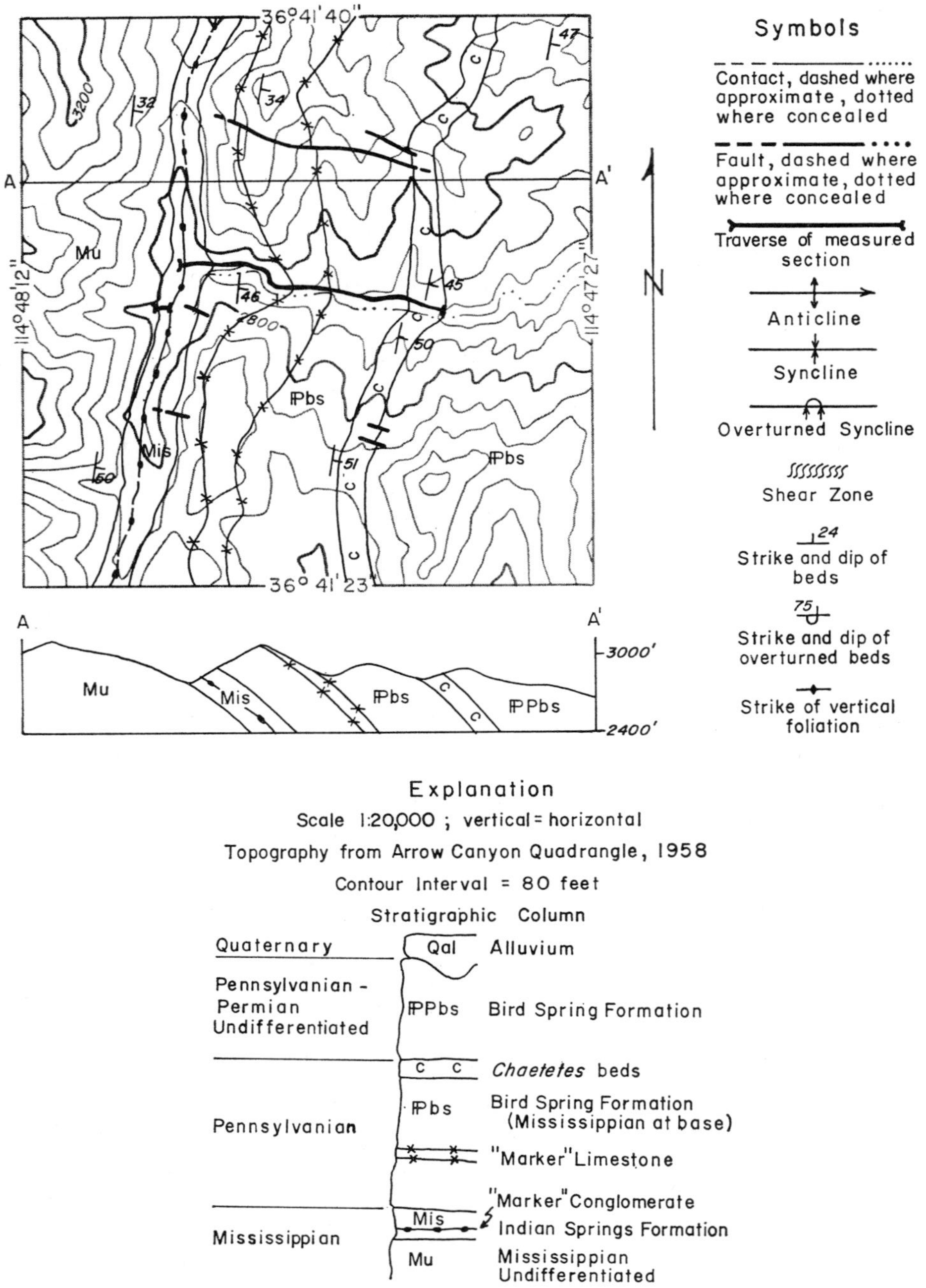

Fig. 21. Geologic map showing line of traverse for Section E, Battleship Wash, Clark County, Nevada.

Unit	*Thickness*	*Lithologic description*
108	4.3	Cover.
107	3.8	Ls, lt to med gry, wea lt bl-gry w/ tan stripe, med bioclas, v f xln mtrx, ool bnd in middle partly silicif, chrt nods in upper 1′, thck bdd, good resist and expos. *Fusulinella* in upper 0.5′.
106	14.8	Cover.
105	2.4	Ls, lt gry, wea same, v f to f bioclas, cptxln mtrx, chrt nods as discont bnds, sl aren, p resist and expos, thn bdd.
104	11.0	Cover, abdt chrt in float.
103	4.3	Ls, lower ½ med gry, upper ½ lt gry, both wea same, f to med bioclas, v f xln mtrx, abdt chrt nods in upper ½, med bdd, fr resist and expos.
102	8.6	Cover, abdt chrt in float.
101	12.1	Ls, med to dk gry, wea lt gry w/ med gry upper 2′, f to med bioclas, v f xln mtrx, few scat chrt nods, dk gry to brn, med bdd, resist, silicif of s/ fos frags.
100	10.0	Cover.
99	10.4	Ls, lt to med gry, wea lt gry, f to med bioclas, cptxln mtrx, abdt chrt nods, conchoidal fract, fr resist and expos.
98	21.5	Ls, lt to med gry, wea same, f to c bioclas, sl arg in middle, cptxln mtrx, thn to med bdd, resist, s/ lamination, cliff former, *Fusulinella* 13′ above base.
97	4.0	Ls, lt gry, wea same (lower 1.4′ wea lt yel-tan), f to med bioclas, v f xln mtrx, med bdd, fr resist and expos, abdt *Chaetetes* heads up to 2′ long, 1′ in diam, silicif in pt, crin cols.
96	13.1	Cover.
95	4.8	Ls, lt to med gry, wea lt gry, f to c bioclas, v f xln mtrx, thn to med bdd (0.1′ to 1.8′), fr expos and resist, *Chaetetes* heads sparse, up to 1.5′ in diam, present at top and bottom, *Syringopora?* top and bottom, *Fusulinella* on top bddg surface, s/ silicif of fos.
94	2.4	Shly ls, lt gry, wea gry-yel, med bioclas, arg mtrx, nonresist, abdt calcite xls from sml fault.
93	5.2	Ls, med gry, wea lt gry, f to c bioclas, v f xln mtrx, med bdd (1′ to 2.2′), resist, fr expos.
92	22.2	Cover.
91	4.9	Ls, lt gry, v f xln, wea lt rust-yel-orng, dnse, conchoidal fract, calcite filled fracts, thck bdd, resist.
90	13.3	Cover.
89	7.4	Ls, med gry, wea lt gry, f to med bioclas, v f xln mtrx, dk gry dolo nods in middle, dk brn to blk chrt nods on upper surface, thck bdd, resist, silicif brach and coral frags.
88	13.0	Cover.
87	10.4	Ls, lt gry, wea gry-tan, f to med bioclas, v f xln mtrx, v aren (f ang well srtd qtz), xbdd (4′ to 5′ long, 1″ to 6″ deep), thn to thck bdd, fr resist and expos.
86	9.4	Ls, lt to med gry, wea lt gry-bl, f bioclas, v f xln mtrx, chrt nods at base and top, stylolites, silicif gastros in chrt nods, thck bdd, grades into overlying unit w/worm? borings, cliff former.
85	13.8	Ls, lt gry, wea same, f to med bioclas, v f xln mtrx, arg in pt, med gry chrt nods in discont bnds w/rust-brn gossans, thn to thck bdd, sb-plnr bddg surfaces, fr resist and expos, silicif brachs.
84	16.4	Cover, float suggest arg chrty ls.
83	14.0	Ls, lt gry, wea lt gry to gry-yel in lower ½, f to med bioclas, v f xln mtrx, abdt dk gry chrt nods in discont bnds, most abdt in lower

Unit	Thickness	Lithologic description
		pt, thck bdd, arg in lower pt, abdt rugose corals in lower pt, *Profusulinella* in upper pt.
82	16.3	Ss, lt gry to gry-yel, wea yel to prpl in upper 6′, lt gry-brn below, calc in lower pt, v f to f grd, ang to sb-ang, well srtd, xbdd, 6″ to 1′ long, 1′ to 2′ deep, thn to med bdd, p resist and expos.
81	11.7	Ls, lt gry, wea yel-gry in lower ½ and lt gry above, v f to med bioclas, dk ferrug-brn wea chrt nods and discont bnds, thn to thck bdd, fr resist and expos, sb-plnr bddg surfaces.
80	3.5	Cover.
79	8.7	Ls, med gry, wea lt gry, f to c bioclas, v f xln mtrx, scat dk gry to blk chrt nods, med bdd, pellets, ool, worm? borings.
78	4.4	Cover.
77	6.7	Ls, med gry, wea lt gry, med to c bioclas, v f xln mtrx, few dk gry to blk chrt nods, sl arg in lower 2′, med bdd, p to fr expos.
76	6.3	Cover.
75	22.7	Ls, lt gry, wea whish at base and grades upward to lt gry, f to med bioclas, arg at base, cptxln mtrx, scat chrt nods and discont bnds, lower 12′ p expos, upper pt ledge former, silicif brachs.
74	6.8	Ls, lt gray wea same, f bioclas, v f xln mtrx, aren, worm? borings, resist, thck bdd.
73	3.2	Ls, med gry, wea lt gry, f to med bioclas, v f xln mtrx, v chrty, chrt wea dk rust-brn, s/ discont bnds, thck bdd.
72	4.2	Ls, lt gry, wea same, f to med bioclas, v f xln mtrx, sl aren, arg at base, stylolites, thck bdd.
71	4.5	Ls, med gry, wea lt gry, few blk chrt nods near base, f bioclas, cptxln mtrx, stylolites, silicif brachs.
70	6.2	Ls, lt gry, wea same, f to med bioclas, v f xln mtrx, sl aren, worm? borings, scat blk to dk gry chrt nods and discont bnds, thck bdd.
69	8.8	Ls, upper 5.4′ lt gry wea same, lower 3.4′ med gry, wea lt gry, f to med bioclas, v f xln mtrx, arg, thn to mass bdd, abdt blk chrt nods up to 3″ in diam and discont bnds.
68	8.6	Ls, med gry, wea lt bl-gry, f to med bioclas, v f xln mtrx, few chrt bnds, wea dk rust-brn, reworking and/or worm? borings, med to thck bdd, fr resist.
67	49.4	Alternating ls and cover, approx 50 percent expos; ls, lt to med gry, wea lt gry, chalk-gry at top, f to c bioclas, finer upward, cptxln to v f xln, mtrx, abdt dk gry to blk chrt nods and discont bnds, even bddg, thn to med bdd, sl aren, fauna sparse, most abdt in lower 5′.
66	5.0	Ls, med gry, wea lt bl-gry, f to med bioclas, v f xln mtrx, arg in lower 1′, mass, resist, calc filled fracts.
65	9.6	Cover.
64	16.7	Ls, lt to med gry, wea lt gry, f to med bioclas, v f xln mtrx, dk brn wea lt gry to blk chrt nods and discont bnds, sl undulty thn to med bdg, fr resist and expos, mass bry, ptly silicif brach frags, s/ worm burrowing and reworking.
63	11.2	Cover.
62	7.7	Ss grades upward into sdy ls (upper 2.9′), lt gry, wea rust-brn lighter upward, f grd, ang, well srtd, scat dk brn wea chrt nods and cannonball size dolo nods, thn to med bdd, fr resist, good expos, worm? burrows, calc mtrx.
61	15.1	Cover.
60	4.2	Ls, lt gry, wea same, med bioclas, v f xln mtrx, dk brn wea chrt nods, lower pt v f bioclas, wea rust-yel w/ sml rdd dk brn to blk chrt nods, sb-plnr bddg, thn bdd, conchoidal fract, nonresist, v p expos.

Unit	*Thickness*	*Lithologic description*
59	10.8	Cover.
58	16.7	Ls, lt gry, wea same, f to med bioclas, v f xln mtrx, sl aren, incg sd content upward, dolo nods and discont bnds, sl arg at base, xbdd, 6′ to 10′ long, 1′ to 2′ deep, thck mass bdd, oolites and/or pellets, cliff former.
57	16.7	Cover.
56	4.3	Sdy ls, lt gry, wea laminated gry-brn, v f to f bioclas, cptxln mtrx, sd grs v f to f ang well srtd qtz, thck bdd, fr resist.
55	1.8	Cover.
54	7.0	Ls, lt gry, wea same, f to c bioclas, v f xln mtrx, blk chrt nods, arg ptgs, resist, thck bdd, fr expos, silicif brach frags.
53	12.8	Ls, lt gry, wea same, v arg, f bioclas, cptxln and clay mtrx, blk chrt nods 1′ to 3′ in diam, plty, p expos.
52	6.8	Ls, lt gry, wea same, f to med bioclas, v f xln mtrx, sl aren, v f ang qtz, med bdd, fr resist and expos, scat dk gry chrt nods, upper ½ incg qtz.
51	5.2	Cover.
50	11.5	Ls, med to dk gry, wea lt gry, cptxln to v f bioclas, cptxln mtrx, abdt gry to blk chrt nods, conchoidal fract, thn to med bdd (0.4′ to 2.0′), fr resist and expos, particularly upper 5′, ferrug gossan on chrt nods, few silicif brach frags.
49	7.8	Cover.
48	14.5	Ls, med gry, wea lt gry, v f to f bioclas, cptxln mtrx, dk gry to blk chrt nod w/ ferrug gossan, thck bdd, resist, p expos, few ptly silicif sol corals in lower 5′
47	65.6	Sdy ls, lt gry, wea lt gry-brn, v f to f bioclas, cptxln mtrx, sd grs v f to f ang well srtd qtz, scat dk brn wea chrt nods and discont bnds, a 2.4′ c bioclas bd 50′ above base, thn bdd in mass unit, fr resist, good expos, calc filled fracts, few *Chonetes* in lower 4′.
46	19.6	Ls, lt to med gry, wea gry-bl, f to c bioclas, v f xln mtrx, aren, scat blk ferrug stnd chrt nods, thck to mass bdd, resist, 2 thin (0.2′ and 1.1′) hash stringers near middle of unit, phaceloid coral biostrome at base, *Syringopora* 2.4′ below top.
45	27.4	Ls, med gry, wea lt gry w/ brn ferrug stnd chrt nods, med to v f bioclas, v f xln mtrx, thn to thck bdd, 0.4′ to 5′, fr resist, *Syringopora* in upper 6′
44	13.9	Ls, lt to med gry, wea lt gry, f to c bioclas, v f xln mtrx, arg ptgs, thn bdd in lower 5′, thck to mass in upper pt, fr resist, pit and cusp wea.
43	3.1	Ls, med gry, wea lt gry-bl, dk blk to brn ferrug stnd chrt nods, undulty med bddg, fr resist, phaceloid biostrome.
42	35.3	Ls, lt to med gry, wea lt gry-tan, f bioclas, v f xln mtrx, abdt dk brn to blk chrt nods in discont bnds, few dolo nods lt tan to dk gry, arg in pt, middle p resist, thn to med bdd, abdt silicif brachs, spiriferids, producs, *Derbyia*? in lower 18′.
41	9.6	Ls, lt to med gry, wea lt gry, f to c bioclas, v f xln mtrx, thck bdd, resist, fos hash in middle, gastros abdt in upper 3′, stylolites.
40	15.7	Sh grading upward into ls: sh, lt gry, plty to bky, wea gry to gry-yel; ls, lt to med gry, wea bl-gry, v f to med bioclas, cytxln mtrx, dk gry chrt nods, fr resist, upper pt p expos; unit p resist.
39	15.4	Ls, dk gry, wea lt gry, v f to f bioclas, cptxln mtrx, abdt blk chrt nods in discont bnds, thn bdd in mass unit, fr resist, abdt *Chonetes*, *Derbyia*?, s/ silicif of fos.
38	14.8	Ls, lt to med gry, wea lt gry to gry-yel in lower half, f to c bioclas,

Unit	*Thickness*	*Lithologic description*
		v f to xln mtrx, dk brn chrt nods in discont bnds, aren (v f ang qtz), sml scale xbdg, thn to thck bdd (0.4″ to 3.0″), undulty bddg, resist cliff former.
37	10.9	Ls, dk gry to blk, wea yel-gry, v f bioclas, cptxln mtrx, petro odor on fresh break, thn discont bddg in mass unit, fr resist.
36	6.2	Calc sh, lt gry, wea gry-yel, spltry to plty, v calc, p resist and expos.
35	13.2	Ls, dk gry, wea lt gry, f bioclas, v f xln mtrx, arg, conchoidal fract, blk chrt nods in upper 4′, thn irreg and discont bddg, rdd mass appearance wea.
34	34.1	Ls, lt to med gry, wea lt gry, to prpl-gry, f to c bioclas, v f xln mtrx, arg in pt, dk gry to blk chrt bnds w/ ferrug gossan, p expos and resist.
33	11.2	Ls, lt to med gry, wea same, f to c bioclas, v f xln mtrx, chrt nod in lower ½ and thn bdd, upper ½ thck bdd to mass.
32	17.7	Ls, med gry wea lt gry, med to c bioclas, v f xln mtrx, lt to dk gry chrt nods in discont bnds, thn to thck bdd, fr resist.
31	7.6	Ls, med to dk gry, f to med bioclas, v f xln mtrx, blk chrt nods in discont bnds, arg in pt, p resist and expos, thn bdd.
30	25.2	Ls, lt to dk gry, wea gry-bl, f to c bioclas, v f to cptxln mtrx, few dk gry chrt nod bnds, petro odor on fresh break, thn to thck bdd, 0.2′ to 3.5′, fr resist, silicif spiriferids, worm? trails on one bddg surface at top, even bddg.
29	39.2	Ls, lt gry, wea same, f to c bioclas, v f xln mtrx, abdt dk gry to blk chrt nods in discont bnds, up to 0.6′ thck, arg, thn to thck bdd (0.2′ to 2.5′), even bdd, fr resist, abdt silicif brachs, s/ sol corals.
28	17.0	Ls, lt gry, wea same, c bioclas, v f xln mtrx, lower 6′ v aren (f ang well srtd qtz), xbdd, upper 11′ w/ abdt fos, med bdd (0.6′ to 2.0′), resist, s/ silicif of fos.
27	2.8	Cover.
26	7.3	Ls, lt gry, wea, same bioclas, v f xln mtrx, sb-plnr bdd surfaces, discont in pt, fr resist, fecal pellets?
25	5.4	Sh, varicolor, lt grn to gry to red to prpl to yel, plty to bky, nonresist, p expos.
24	20.5	Ls, as in underlying unit w/ fewer dolo nods, spiriferids restricted to lower 2′, producs from arg bds.
23	12.5	Ls, lt gry, wea bl-gry, bioclas, v f xln mtrx, dk gry-bl dolo nods in discont bnds, thn bdd in thck units, resist, abdt silicif brachs, arg ptgs.
22	10.2	Ls, lt gry, wea gry-bl, bioclas, cptxln to v f xln mtrx, sl aren in lower 4′ and wea tan, arg in upper 3′, dk gry chrt nods wea ferrug-brn, thn to thck bdd, fr resist, abdt fos in upper 3′, s/ silicif of fos.
	———	
	1,098.1	Total thickness of measured part of Bird Spring formation; conformable with Indian Springs formation.
		Indian Springs formation
21	7.3	Cover.
20	2.4	Ls, lt gry, wea rusty-red-gry, bioclas, v f xln mtrx, arg ptgs, v ferrug on wea surface, thn bdd, p expos, abdt spiriferids on upper bddg surface.
19	13.6	Cover.
18	5.7	Ss, lt yel to gry, wea tan to rust-brn, v f to f grd, ang to sb-rdd, fr srtg, ferrug and calc cmt, xbdd, current ripple marks, thn bdd, p expos.
17	12.3	Ls, lt gry to yel, wea yel-gry to orng, v f xln to f bioclas, ferrug

Unit	*Thickness*	*Lithologic description*
		stng and mtrx in pt, cptxln mtrx in pt, arg ptgs, thn to med bdd, bddg wavy to even (0.2′ to 1.9′), abdt producs in lower 8′.
16	13.9	Cover.
15	0.8	Ls, gry-orng, wea ferrug-rust-orng, c bioclas, ferrug mtrx, sft, even bdg.
14	2.4	Ss, lt prpl-gry, wea dk prpl-gry, v f to f grd, ang, well srtd, calc cmt, thn bdd, fr resist.
13	50.2	Cover.
12	0.4	Ss, wh to lt gry, wea yel-brn, v f grd, ang, well srtd, calc cmt, even thn bdd, unidentified blk specks.
11	2.2	Sh, dk gry to blk, wea same, spltry to plty, nonresist, p expos.
10	5.5	Ls, lt gry, wea gry-tan, med bioclas, cptxln mtrx, thck bdd, upper 2′ cgl, reddish-gry, wea dk reddish-prpl to ferrug-orng, clasts up to 2′ in diam, ferrug mtrx, resist.
9	24.7	Cover except for a 1′ thck ls as in the underlying unit 10′ above base.
8	1.0	Ls, gry-orng, wea rust-orng, med to c bioclas, f xln mtrx and ferrug mtrx, thn bdd, p resist.
7	26.8	Cover.
6	0.7	Ls, gry-yel, wea rust-orng, f bioclas to c, v f xln mtrx and ferrug, med bdd.
5	28.4	Cover.
4	2.5	Sh, varicolor, prpl to yel to red, bky to plty, nonresist, p expos.
3	7.1	Ss, lt yel, wea yel-orng, v f grd, ang, well srtd, sml festoon xbds, ripple marks (oscillation 2″ length, 1.4″ depth, N 30° E along crests), grades upward into lt yel c bioclas ls, thn to med bdd, arg in pt, p expos.
2	4.3	Sh, varicolor, prpl to yel to red, bky, nonresist, grades upward into ss, p expos.
	212.2	Total thickness of Indian Springs formation; conformable with Battleship Wash formation.
		Battleship Wash formation
1	Not measured	Ls, lt to med gry, wea lt gry-bl, bioclas, f xln mtrx, calcite-filled fracts, thck bdd, resist, cliff former.

SECTION F. LAS VEGAS RANGE

Location.—Lat. 36° 39′ 29″ N, Long. 114° 57′ 26″ W to Lat. 36° 40′ 42″ N, Long. 114° 57′ 41″ W.

Comments.—Section measured on north dipping series of *en échelon* fault blocks (fig. 22). Indian Springs formation poorly exposed, considerably contorted, and cut by numerous small faults; Bird Spring formation well exposed. Line of section offset several times to avoid faults, moved down dip or along strike to obtain better exposure. UCLA locality 5247.

Unit	*Thickness*	*Lithologic description*
		Bird Spring formation
90	16.1	Ls, lt to med gry, wea lt gry, f to med bioclas, v f xln mtrx, stylolites, thn to med bdd, lower pt sl arg and v chrty, cliff former, *Chaetetes* heads up to 2′ in diam, fr abdt, s/ silicif of brachs.
89	27.3	Approx 85 percent cover; ls, lt gry, wea med gry, med bioclas, med bdd, p expos, silicif brach frags.

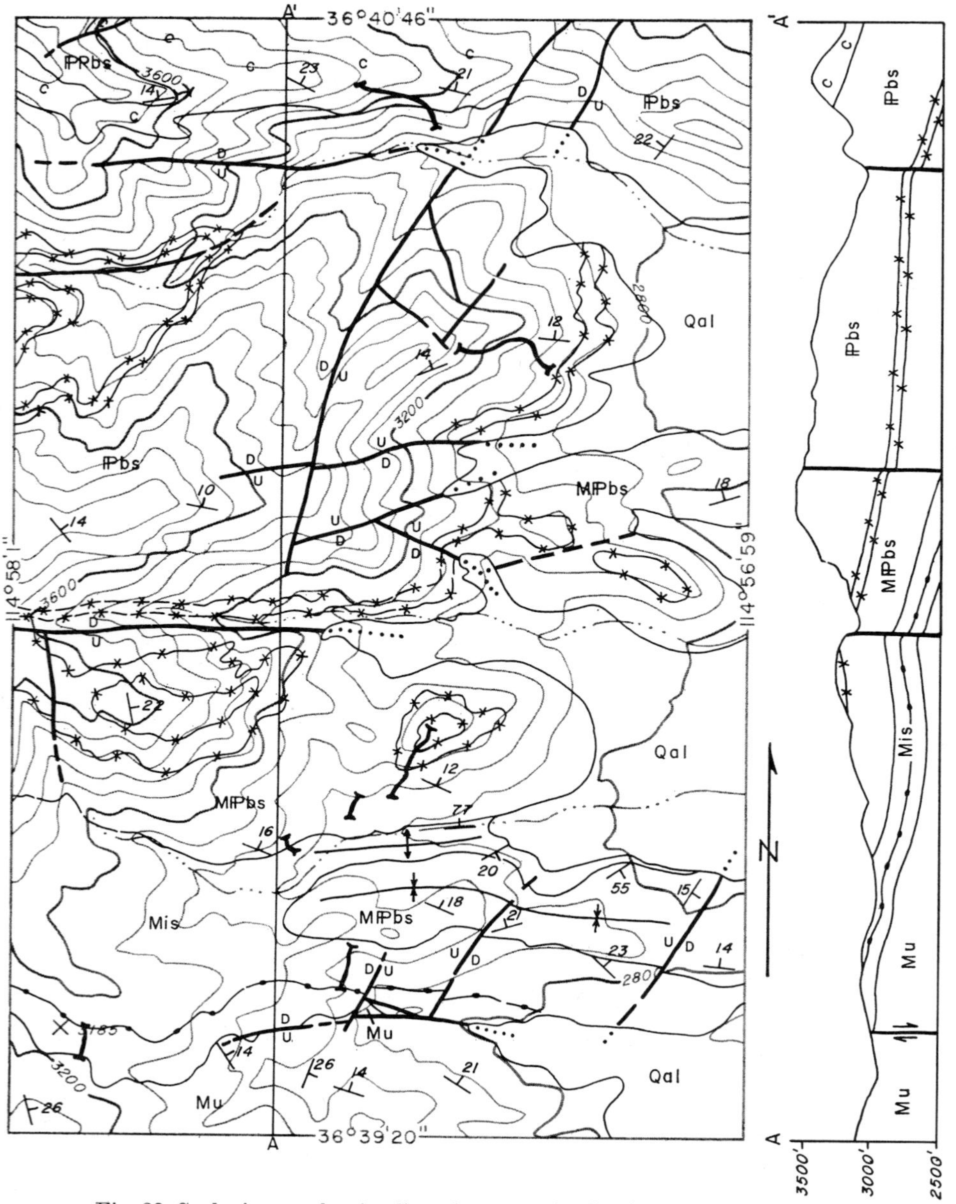

Fig. 22. Geologic map showing line of traverse for Section F, Las Vegas Range, Clark County, Nevada. See figure 21 for explanation of symbols.

Unit	*Thickness*	*Lithologic description*
88	1.0	Aren ls, lt gry, wea gry-tan, med bioclas, abdt f ang qtz, v f xln mtrx, plty wea.
87	4.7	Ls, lt gry, wea same, f to med bioclas, med to thck bdd, silicif *Chaetetes* heads up to 2′ long and 1.4′ in diam.
86	4.5	Ls, lt gry, wea tan-gry, f to med bioclas, aren, thn bdd, chrt nods, p resist.

Unit	*Thickness*	*Lithologic description*
85	2.6	Ls, lt gry, wea same, f to med bioclas, med to thck bdd, brn chrt nods, silicif *Chaetetes* heads up to 2′ long and 1.5′ in diam.
84	16.0	Cover.
83	1.1	Ls, lt gry, wea same, f to med bioclas, v f xln mtrx, dk grn wea chrt nods, med bdd, fr resist and expos
82	13.0	Cover.
81	7.0	Ls, lt to med gry, wea yel-gry in lower 3′ and lt gry w/ brn dolo bnds above, f to med bioclas, v f xln mtrx, med to thck bdd, resist, abdt fos frags, crin cols.
80	15.6	Cover.
79	24.5	Interbdd ls and arg ls; ls, med gry, wea gry-bl, sl aren, f to med bioclas, cptxln mtrx, scat large (up to 1′ diam) chrt nods, med to thck bdd, resist, arg ls, dk gry, wea lt gry, f to c bioclas, v f xln mtrx, chrt nods in upper pt, p resist, good expos, v abdt bry in upper unit.
78	13.8	Ls, med gry, wea lt gry, med bioclas, v f xln mtrx, few large (up to 1′ diam) chrt nods, mass bdd, resist, cliff former, brn silicif *Chaetetes* heads up to 2.4′ in diam in lower 5.4′, sml and rare in upper 8.4′.
77	15.5	Cover.
76	2.1	Ls, med gry, wea lt gry, f to med bioclas, v f xln mtrx, thn to med bdd, p resist and expos.
75	16.8	Cover.
74	20.3	Ls, lt to med gry, wea lt gry w/ brn chrt bnds in upper 13.3′, lt yel-tan in lower 7′, ft to c bioclas, v f xln mtrx, abdt fos hash, arg in middle, thn to med bdd, p to fr resist and expos, s/ silicif of fos.
73	16.1	Cover.
72	11.3	Ls, med gry, wea lt gry w/ olive cast, f to med bioclas, abdt brn to blk chrt nods up to 4″ thck, 2′ long, incg chrt upward, thn to med bdd, crin cup at top.
71	7.4	Ls, dk gry, wea bl-gry, f to med bioclas, v f xln mtrx, thck bdd, resist, cliff former, *Chaetetes* heads (0.4′ to 1.0′ in diam) in upper 3′, s/ heads 1.5′ long.
70	15.1	Ls, med to dk gry, wea yel-brn in upper 6′, and gry-brn below, v chrty, sl slty in upper 6′, v f bioclas, cptxln mtrx, conchoidal fract, thn bdd in mass units.
69	6.0	Calc sh, lt gry, wea wh-gry, crumbly, chlky, blk chrt nods, p resist and expos.
68	7.8	Ls, med gry, wea gry-bl, f to med bioclas, v f xln mtrx, thn to med bdd, fr expos, p resist, silicif brach frags in lower 2′, silicif *Syringopora* in middle.
67	45.4	Ls, med to dk gry, wea gry-brn in lower 35′ and gry-bl in upper 9.4′ f to med bioclas, v f xln mtrx, abdt dk gry to brn to blk chrt nods and discont bnds, stylolites, resist, cliff former, silicif brach frags.
66	5.4	Cover.
65	13.2	Interbdd, calc sh and ls: calc sh, lt gry, wea wh to gry-yel, blk chrt nods, plty to bky; ls, lt gry, wea same, med to c bioclas, f xln mtrx, blk chrt nods, thn bdd; unit p resist and expos.
64	6.2	Cover.
63	21.9	Ls, dk gry to blk, wea gry-tan, grades upward from v f to c bioclas w/ s/ lt gry pbbl-sized ls clasts, v f to cptxln mtrx, dcrg chrt upward, w/ nodular bnds up to 1′ thck 1′ below top of unit, lower pt thn bdd in mass unit, upper pt thck to mass bdd, resist cliff former.

Unit	Thickness	Lithologic description
62	14.2	Cover.
61	8.1	Ls, lt gry, wea gry-tan, med bioclas, v f xln mtrx, v aren (f ang well srtd qtz), xbdd in lower 2′, appears laminated on wea surface, ool, fr expos.
60	30.8	Ls, lt to med gry, wea gry-tan to gry-bl, f to c bioclas, v f xln mtrx, sl aren at base, dk gry to brn chrts nods and discont bnds, no chrt in upper 3′, med to mass bdd, v resist cliff former.
59	16.8	Ls, lt to med gry, wea lt cream-gry in lower 6′ and lt gry above, f to med bioclas, cptxln mtrx, dk gry to blk chrt nods up to 4″ diam, thn bdd upper pt, mass lower, p resist and expos.
58	16.3	Ls, lt to med gry, wea lt gry and cream-gry, f to med bioclas, v f xln mtrx, v chrty bds wea cream-gry, dk gry to brn chrt nods and discont bnds, med bdd, fr resist and expos, few fos.
57	12.7	Cover.
56	30.4	Ls, med gry, wea gry-bl, f to med bioclas, v f xln mtrx, abdt brn to dk gry chrt nods and discont bnds, med to mass bdd.
55	10.2	Ls, lt gry, wea cream-gry-yel, f bioclas, cptxln mtrx, slty and aren, v f ang qtz, dk gry chrt nods, calcite-filled nods, nodular wea, v conspicuous bnd owing to wea color, mass w/ med bddg as discont bnds.
54	4.3	Sdy ls, lt gry, wea same, f to med bioclas, v aren (f ang qtz), v f xln mtrx, xbdd (0.5′ to 2.0′ long, 0.2′ to 0.6′ deep), thck bdd, fr resist.
53	17.8	Ls, lt gry, wea lt gry to gry-brn, f to med bioclas, aren, v f xln mtrx, dk brn to gry chrt nods and discont bnds, s/ chrt looks like slump and flow structs, fr resist and expos, arg ptgs, s/ silicif of fos, med bdd.
52	6.1	Ls, lt gry, wea same, f to med bioclas, v f xln mtrx, sl aren in upper 1′, thn to med bdd, fr resist, ledge former, arg ptgs.
51	26.5	Approx 45 percent expos; ls, lt to med gry, wea lt gry, f to med bioclas, v f xln mtrx, gry-brn chrt nods, thn to med bdd (0.2′ to 1.3′), p resist, slope former.
50	4.5	Ls, med gry, wea lt gry, med bioclas, v f xln mtrx, appears laminated on wea surface, gry-brn chrt nods in base, mass, cliff former.
49	14.2	Ls, lt gry, wea same f to med bioclas, v f xln mtrx, sl aren (slt to v f ang qtz), abdt dk gry chrt nods and discont bnds which wea dk brn, fr resist, thn to med bdd.
48	5.2	Calc sh, lt gry, wea gry-yel, plty, pt bky, blk chrt bnds, nonresist, p expos.
47	7.5	Cover.
46	13.3	Ls, lt to med gry, wea lt gry to gry-tan, f to c bioclas, v f xln mtrx, sl aren, scat blk chrt nods, arg, ptgs, med bdd, p resist and expos, crin cols, spiriferids and gastros in upper 2′.
45	5.8	Cover, float suggests v chrty arg ls.
44	6.1	Ls, lt gry, wea lt gry-bl, f to c bioclas, v f xln mtrx, arg ptgs, thn to med bdd, sft, p resist, fr expos.
43	6.8	Cover.
42	23.8	Gry to brn to blk chrt with ferrug gossan in ls mtrx, 80 percent chrt; ls, lt gry-brn, wea gry-tan, v f xln to cptxln, calcite-filled vugs, geodes, thn bdd, p resist, fr expos.
41	6.8	Ls, lt to med gry, wea lt gry, f to med bioclas, v f xln mtrx, lt gry to brn chrt nods w/ ferrug gossan, sl aren in upper 2′, arg ptgs, thn to med bdd, cliff former, silicif brach frags in lower 3′.

Unit	*Thickness*	*Lithologic description*
40	8.7	Cover.
39	24.3	Ls, med gry to blk, wea lt gry-yel, v f to f bioclas, cptxln mtrx, strong petrol odor on fresh break, abdt dk gry to blk chrts nods up to 3″ in diam and discont bnds, thn bdd, p resist, fr expos.
38	15.1	Cover.
37	20.0	Ls, lt to med gry, wea lt gry to bl-gry, f bioclas, cptxln mtrx, stylolites, alternating sl chrty bds and v chrty bds, dk brn chrt nods and discont bnds, thn to mass bdd, worm? borings, resist, ledge former.
36	11.4	Ls, lt gry, wea gry-brn, f to med bioclas, v f xln mtrx, sl aren, dk brn to blk chrt nods and thn discont bnds, thn to med bdd (0.2′ to 1.3′), appears laminated on wea surface, fr expos, nonresist, slope former.
35	40.2	Cover.
34	12.7	Ls, med gry, wea bl-gry, med to thck bdd (1.1′ to 4.3′), f to c bioclas, v f xln mtrx, gry to yel-brn chrt nods and discont bnds, large dolo nods in upper 3′ up to 4′ long, 3′ wide, 2′ thck, dolo nods have ferrug gossan, chrty gossan locally, resist cap for big cliff.
33	66.9	Ls, lt to med gry, wea lt brn, wea as a big brn cliff from a dist, v f to f bioclas, cptxln mtrx, v chrty in nods and discont bnds, nods up to 4″ diam, bnds from ¼″ to 6″ thck, v aren, slt and v f grd qtz, thn bdd in a mass unit.
32	34.0	Ls, med to dk gry, wea lt to med gry to tan-orng, v f bioclas, cptxln mtrx, v slty and aren (v f qtz), aren bds wea tan-orng, abdt dk brn to blk chrt nods and discont bnds, dolo cores in s/ chrt nods.
31	7.4	Cover.
30	36.3	Ls, as in underlying unit except slope former.
29	57.4	Ls, med gry to blk, wea gry to lt prpl-gry, upper 35′ resist ledge former, rest slope former, abdt dk gry to brn to blk chrt nods, ⅛″ to 4″ in diam, numerous discont bnds, dolo cores in s/ chrt nods, plty wea in pt, s/ pit and cusp wea on med gry bds, thn to med bdd.
28	59.6	Ls, dk gry to blk, wea lt to med gry, v f to c bioclas, pbbly surfaces in lower 10′, v f xln mtrx, arg, arg ptgs, dk brn to blk chrt nods and discont bnds, approx 65 percent expos, thn to thck bdd (0.1′ to 3′), v arg bds wea w/ prpl cast, plty wea, middle of unit wea lt gry, abdt fos, slope holder.
27	31.4	Cover.
26	6.0	Ls, dk gry to blk, wea orng-tan grading upward into lt gry, f to c bioclas, v f xln mtrx, med to thck bdd, arg ptgs, dk brn chrt nods, resist ledge former.
25	12.9	Ls, dk gry to blk, wea prpl-gry to yel-gry, f bioclas, v f xln mtrx, arg in pt, thn to med bdd, few thn interbds of lt gry plty calc sh, nonresist.
24	43.2	Alternating cover (65 percent) and ls, med to dk gry, wea gry-bl, f to med bioclas, cptxln to v f xln mtrx, sl arg, dk gry to blk chrt nods and discont bnds, thn to med bdd, abdt fos, p resist and expos.
23	20.5	Ls, med gry, wea lt bl-gry, interbdd f to c bioclas, v f xln mtrx, dk brn to blk chrt nods and discont bnds, few dolo nods, thn to med bdd (0.4′ to 2′), arg ptgs, sl aren at base and top of unit.
22	5.6	Ss grading up into sdy ls, lt gry to tan, wea gry-tan, med bioclas, v f xln mtrx, sd grs f ang well srtd, calc mtrx in ss, few chrt nods in upper 1.5′, med bdd, fr resist and expos.

Unit	*Thickness*	*Lithologic description*
21	3.5	Sh, varicolor grn to yel to red, plty, calc, nonresist, p expos.
20	6.9	Ls, lt gry, wea same, c bioclas, v f xln mtrx, sl fri on wea surface, med to thck bdd, resist, good expos, abdt spiriferids particularly in upper 2′.
19	3.2	Cover.
18	34.5	Ls, lt to med gry, wea lt bl-gry, med to c bioclas, v f xln mtrx, arg ptgs, dk brn chrt nods, thn to thck bdd, 0.3′ to 3.0′, s/ silicif of fos, ramose bry, bellerophontid gastros in upper 1.0′, upper 8′ v resist, otherwise p resist, fr expos.
17	5.0	Ls, lt to med gry, f to c bioclas, v f xln mtrx, wea prpl-gry, dk gry chrt nods and discont bnds in middle of unit, chrt wea dk ferrug brn, thn to thck bdd, fr resist and expos.
	1,215.6	Total thickness of measured part of Bird Spring formation; conformable contact with Indian Springs formation.
		Indian Springs formation
16	17.6	Cover 85 percent; ls, lt olive-gry, wea yel-gry, med to c bioclas, v f xln mtrx, med bdd, p resist and expos.
15	2.5	Ls, olive-gry to prpl, wea yel-gry to olive-gry, f to c bioclas, v f xln mtrx, ferrug gossan w/ chrt pbbl on upper bddg surface, bddg 0.1′ to 1.3′, fr resist and expos, abdt fos.
14	12.9	Cover.
13	2.8	Ss, sb-qtzite, gry-grn to prpl, wea dk ferrug brn, v f to f grd, ang, well srtd, sml festoons (0.1′ to 0.2′ long, 0.02′ to 0.05′ deep), mass, fr resist and expos.
12	12.0	Cover.
11	8.4	Ss, lt tan, wea lt tan to rust-brn, f to c grd, sb-ang to sb-rdd, poorly srtd, calc cmt, fri, xbdd (0.3′ to 1.0′ long, 0.05′ to 0.15′ deep), thn to thck bdd, p resist and expos, plant frags.
10	27.0	Sdy ls, lt gry to gry-brn, wea rust-tan to orng-tan, c bioclas, v f xln mtrx, v aren (med to c grd well rdd qtz), xbdd (0.5′ to 1′ long, 0.1′ to 0.3′ deep), thck bdd w/ thn discont bddg planes in thck units, grades into cgl in upper 7′, fr resist and expos, v fos, considerable attrition of fos.
9	40.0	Cover.
8	3.5	Ls, lt gry-red, wea orng-red w/ dk red 0.5′ cgl at top, f to c bioclas, v f xln mtrx, med bdd, resist, particularly cgl, v ferrug.
7	14.1	Cover.
6	2.3	Ls, med gry, wea rust-red, dk ferrug gossan on top bddg surface, med to c bioclas, v f xln and ferrug mtrx, thck bdd, fr resist, p expos.
5	96.7	Cover, float and animal diggings suggest this interval is mostly blk plty sh (est 95 percent), reddish-brn wea bioclas ls (3 percent), and brn wea ss (2 percent); interval a slope former.
4	1.0	Ss, med to dk gry, wea brn to gry, v f grd, ang, well srtd, sl calc cmt, sb-qtzite, med bdd, p expos.
3	5.0	Cover.
2	4.0	Ss, wh, wea lt gry, f grd, ang, well srtd, thn bdd, p expos, 25 percent expos.
	249.8	Total thickness of Indian Springs formation; conformable with Battleship Wash formation.
		Battleship Wash formation
1	Not measured	Ls, lt gry, wea olive-gry to gry-bl, med to c bioclas, v f xln mtrx, med bdd, undulty bddg, resist, cliff former.

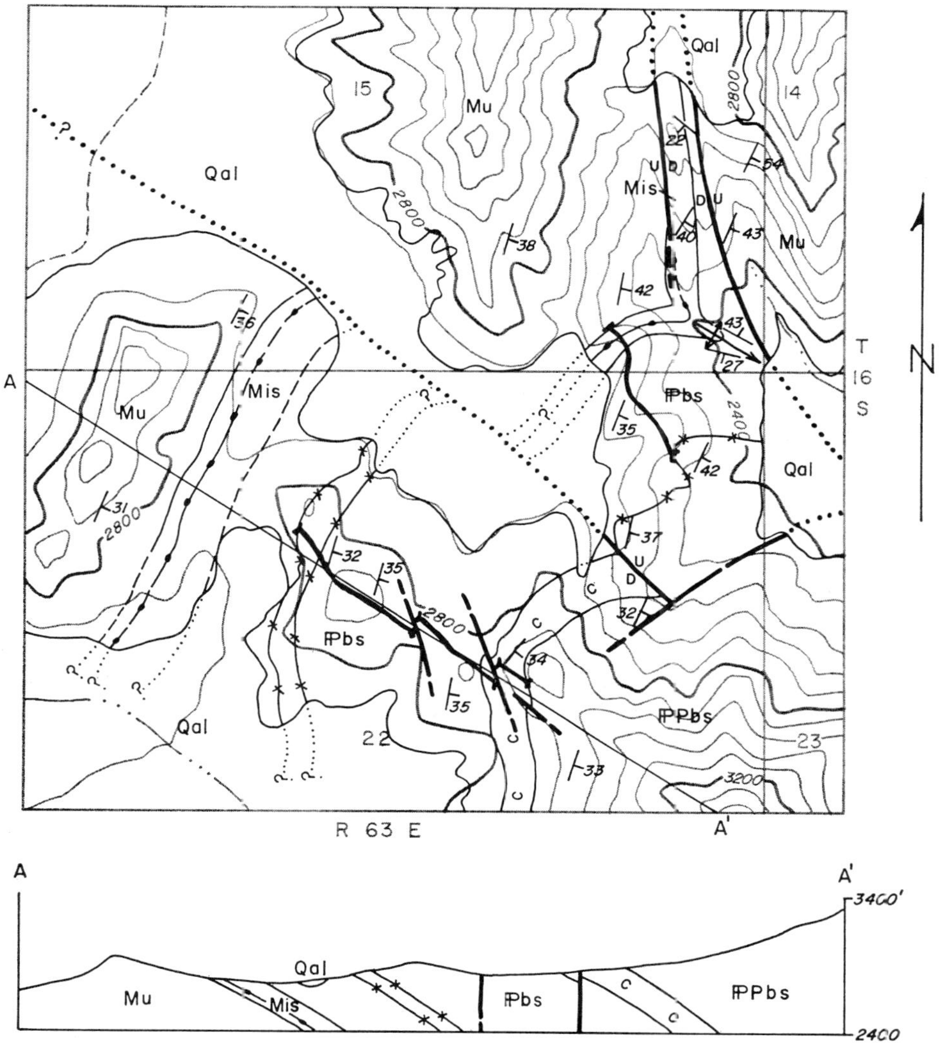

Fig. 23. Geologic map showing line of traverse for Section G, Dry Lake, Clark County, Nevada. See figure 21 for explanation of symbols.

SECTION G. DRY LAKE

Location.—SE 1/4, SE 1/4 sec. 15 and SE 1/4, NW 1/4 to SW 1/4, NE 1/4 sec. 22, T. 16 S., R. 63 E., Clark County, Nevada.

Comments.—Section measured in two segments (fig. 23). Lower 487.1′ measured in northern segment. Exposures good to excellent; minor contortion in Indian Springs formation. Upper 934.6′ measured along ridge crest 1/8 mile southwest of northern segment. Numerous small faults in saddles along ridge crest required several offsets. Exposure of Bird Spring formation along ridge crest poor to fair. UCLA locality 5248.

Unit	Thickness	Lithologic description
		Bird Spring formation
91	16.0	Ls, lt to med gry, wea lt gry, f to med bioclas, v f xln mtrx, med to thck bdd, resist cliff former, chrt nods w/ ferrug gossan, v abdt *Chaetetes* heads, cylindrical and spherical, up to 2′ long, cylindrical heads most common.
90	9.1	Cover.
89	16.2	Ls, lt gry, wea same, f to med bioclas, v f xln mtrx, gry to brn chrt nods and discont bnds, thn to med bdd, fr resist and expos, s/ silicif of fos.
88	48.4	Cover.
87	53.2	Ls, lt gry, wea same, f bioclas, cptxln to v f xln mtrx, thn to med bdd, p resist, 50 percent expos, plty wea.
86	6.1	Ls, med gry, wea bl-gry, med to c bioclas, v f xln mtrx, thn to med bdd, resist cliff former.
85	5.4	Cover.
84	8.4	Ls, lt to med gry, wea lt gry, f to med bioclas, v f xln mtrx, med bdd, fr resist and expos, s/ silicif of fos, fr abdt *Chaetetes* heads up to 1.8′ in diam.
83	7.2	Cover.
82	15.8	Ls, lt to dk gry, f to c bioclas, v f xln mtrx, wea lt gry, med bdd, good resist and expos, v abdt *Chaetetes* heads up to 2.5′ in diam, s/ silicif of fos, *Fusulinella* on uppermost bddg surface.
81	11.0	Cover.
80	18.8	Ls, lt to med gry, wea lt gry, f to med bioclas, v f xln mtrx, abdt gry to brn chrt nods, thn to med bdd, fr resist, 60 percent expos, *Profusulinella* at base.
79	19.0	Cover.
78	15.0	Ls, lt to med gry, wea same, f to med bioclas, v f xln mtrx, abdt chrt in upper 5′, med bdd, fr resist and expos, s/ silicif of fos.
77	13.8	Cover.
76	3.2	Ls, med gry, wea lt gry, f bioclas, cptxln mtrx, dk brn chrt nods, thn to med bdd, fr resist and expos.
75	3.8	Cover.
74	9.1	Ls, lt gry, wea bl-gry, f to med bioclas, v f xln mtrx, med bdd, resist cliff former.
73	9.0	Ls, lt gry, wea gry-brn, f to med bioclas, v f xln mtrx, aren (f ang qtz), abdt gry to brn chrt, med bdd, fr resist and expos.
72	8.2	Cover.
71	2.0	Ls, lt gry, wea gry-tan, f bioclas, cptxln mtrx, v aren (f ang qtz), med bdd, p expos.
70	27.8	Cover.
69	10.8	Ls, lt gry, wea same, f to med bioclas, v f xln mtrx, gry to brn chrt nods in lower ½, s/ reworking and burrowing? in upper ½, med bdd, fr resist and expos.
68	16.2	Cover.
67	43.7	Ls, lt gry, wea same, f to med bioclas, aren in lower pt (f ang qtz), scat gry to brn chrt nods, most abdt in aren pt, v f xln mtrx, wea wh to off wh at base, med bdd, 75 percent expos, *Profusulinella*? above base of unit.
66	22.7	Cover.
65	16.0	Ls, lt to med gry, wea bl-gry, f to med bioclas, v f xln mtrx, gry to brn chrt nods, silicif nods up to 1.5′ diam, thn to med bdd, 70 percent expos.

Unit	*Thickness*	*Lithologic description*
64	31.2	Cover.
63	34.6	Ls, lt gry, wea bl-gry, f to med bioclas, v f xln mtrx, aren in pt, gry to brn chrt nods in few bds, med bdd, 80 percent expos.
62	2.5	Ls, lt gry, wea gry-tan, f bioclas, cptxln mtrx, aren (f ang qtz), few chrt nods, fr resist and expos.
61	7.5	Cover.
60	6.5	Ls, lt gry, wea lt bl-gry, f bioclas, v f xln mtrx, med bdd, fr resist and expos, silicif brach? frags.
59	8.6	Cover.
58	11.5	Ls, lt gry, wea same, f to med bioclas, v f xln mtrx, aren in pt, thn to med bdd, fr resist and expos.
57	10.6	Cover.
56	35.0	Ls, lt to med gry, wea lt bl-gry, f to med bioclas, v f xln mtrx, chrt nods and discont bnds, silicif nods up to 2′ in diam, reworking and worm borings in basal part, thn to thck bdd, s/ silicif of fos.
55	66.1	Ls, lt gry, wea same, uppermost 2.3′ wea wh, f bioclas, cptxln to v f xln mtrx, sl aren (f ang qtz), gry to brn chrt nods, med bdd, p expos, est 40 percent expos, abdt chrt in float.
54	17.1	Cover.
53	14.8	Ls, lt gry, wea same, f bioclas, cptxln to v f xln mtrx, aren (f ang qtz), gry to brn chrt nods, med bdd, fr resist and expos.
52	13.7	Ls interbdd w/ blk chrt nodular bnds, ls, wh to lt gry, wea wh to off wh, v chlky, grades upward into f bioclas, sft, nonresist, p expos.
51	15.3	Cover.
50	8.8	Ls, lt gry, wea same, f bioclas, aren, v f xln to cptxln mtrx, gry to brn chrt nods, med bdd, upper 2.4′ xbdd, long foresets, fr resist and expos.
49	35.4	Cover.
48	6.8	Ls, med gry, wea lt gry, f to c bioclas, aren in lower pt, v f xln mtrx, thn to med bdd, fr resist and expos.
47	20.0	Ls, lt gry, wea bl-gry, f to med bioclas, v f xln mtrx, med bdd, stylolites in upper pt, middle 10′ cover, fr resist and expos.
46	37.8	Cover.
45	8.0	Ls, med to dk gry, wea med bl-gry, f to med bioclas, v f xln mtrx, lrg silicif producs up to 4″ across and 1″ thck, resist, well exposed, silicif mollusks, thck bdd.
44	106.9	Ls, lt gry, wea lt gry to med gry to gry-brn, from a dist looks rust-brn, v aren (v f ang well srtd qtz), abdt chrt nods (lt gry to brn), f bioclas, cptxln to v f xln mtrx, fr resist and expos, thn bdd in a mass unit, bddg indistinct from any dist.
43	13.5	Sh, lt gry to yel to prpl, wea lt yel at base, grades up into prpl and back to yel at top, calc, plty to bky, sl micac, p resist, fr expos locally. Offset to west approximately 300 yards.
42	95.4	Ls, med to dk gry, wea lt gry, v f to f bioclas, sl arg, cptxln mtrx, numerous gry and brn chrt nods, s/ discont bnds, p resist, fr expos, lower 45′ contains interbdd med gry c bioclas thck to med bdd ls interbdd with ls described for this interval, s/ silicif of fos.
41	83.8	Ls, dk gry to blk, wea lt gry to prpl-gry, f bioclas, arg, cptxln mtrx, scat chrt nods, ferrug nods ⅛″ to ½″ in diam locally, strong petro odor on fresh break, thn bdd, wea plty, s/ silicif of fos, p expos, upper ½ cover and lower ½ 40 percent cover.
40	23.8	Ls, lt to med gry, wea gry-bl, med bioclas, v f xln mtrx, med bdd, resist, well exposed, silicif fos in pt.

Unit	*Thickness*	*Lithologic description*
39	6.8	Sdy ls, lt gry, wea gry-tan, f to med bioclas, v f xln mtrx, sd f grd ang well srtd, thn to med bdd, fr resist and expos, s/ silicif of fos.
38	6.6	Cover.
37	27.6	Ls, lt gry, wea lt to med gry-bl, med to c bioclas, v f xln mtrx, scat gry to brn chrt nods, thn to med bdd, fr resist and expos, silicif fauna in pt.
36	14.3	Ls, lt gry, wea bl-gry, med to c bioclas, v f xln mtrx, dk gry to tan chrt in discont bnds, thn to med bdd, resist, cliff former, silicif fos.
35	3.3	Cover.
34	12.7	Ls, lt gry, wea lt gry to olive-gry, c bioclas, v f xln mtrx, thn to med bdd, sl aren at base, s/ silicif fos, good resist and expos.
	1,222.4	Total thickness of measured part of Bird Spring formation; conformable contact with Indian Springs formation.
		Indian Springs formation
33	6.0	Cover.
32	2.3	Ls, lt gry, wea olive-gry, w/ ferrug coat on uppermost bddg surface, c bioclas, v f xln mtrx, thn bdd, fr resist and expos, abdt *Spirifer?*, few ramose bry, sol corals, crin cols.
31	11.6	Cover.
30	3.0	Ls, lt gry, tan, wea lt gry to olive-tan, f to c bioclas, med bdd, v f xln mtrx, sl reddish wea on uppermost bddg surface, fr resist and expos.
29	4.5	Cover.
28	0.3	Ls, gry-tan, wea olive-tan to rust-red, med bioclas, v f xln mtrx, thn bdd, p expos.
27	4.7	Cover.
26	2.0	Ls, lt gry, wea olive-tan, f to med bioclas, v f xln mtrx, ferrug gossans on upper 2 bddg surfaces, chrty on uppermost surface, thn bdd, fr resist and expos.
25	7.2	Cover.
24	2.6	Ls, lt gry, wea olive-gry to gry-tan, c bioclas, v f xln mtrx, ferrug gossan on upper bddg surface, thn to med bdd, fr resist and expos.
23	12.0	Ls, lt gry to gry-yel, wea rust-orng at base and grades up into yel-gry, f to c bioclas, v f xln mtrx, thn to med bdd, fr resist and expos.
22	4.1	Cover.
21	2.0	Ss, lt gry, wea lt gry to tan, v f grd, sb-qtzitic, ang, well srtd, micac, thn bdd, fr resist, p expos.
20	9.0	Cover.
19	3.2	Ls, yel-tan, wea orng-brn to red-brn, med to c bioclas, thn to med bdd, fr resist, cover in pt.
18	14.6	Sh, gry to blk, wea lt gry, plty, nonresist, p expos, upper ½ mostly cover.
17	5.9	Ls, gry-grn, wea gry-tan to yel-brn, med to c bioclas, v f xln mtrx, arg interbds.
16	32.7	Sh, lt to dk gry, wea lt gry to yel-gry to prpl, calc in pt, plty, brittle, nonresist, abdt fos, p to fr expos, 15 percent cover.
15	2.2	Cgl, lt gry-brn to dk gry, wea lt brn, c bioclas up to pbbls, med bdd, f xln mtrx, p resist and expos.
14	5.8	Cover.
13	15.4	Ss, lt gry to brn, wea tan to lt gry, v f to f grd, ang, well srtd, sml scale xbdd, thn to med bdd, p resist and expos, s/ sml folding.

Unit	*Thickness*	*Lithologic description*
12	2.6	Cover.
11	4.3	Ls, olive-gry, wea same to reddish-gry, 0.6′ cgl wea dk red-brn on top, f to med bioclas, v f xln mtrx, ferrug mtrx in pt, pbbls up to 2″ diam in cgl, ferrug and silicif pbbls and nods, fr resist and expos, cgl contains gastros, pelecys, conodonts, producs, *Spirifer?*
10	12.2	Mostly cover; sh, lt to med gry, wea yel-gry-grn, plty, nonresist, p expos.
9	1.4	Ls, gry-yel to reddish-gry, wea yel-gry, med bioclas, ferrug and v f xln mtrx, dk red-brn gossan on upper bddg surface, med bdd, p resist and expos, producs, crin cols, *Spirifer?*.
8	11.8	Cover.
7	0.9	Ls, lt gry-tan, wea gry-yel, f to med bioclas, v f xln mtrx, thn bdd, p resist and expos.
6	0.5	Ss, tan, wea orng-gry, v f grd, ang, sb-qtzitic, thn bdd, p resist and expos.
5	2.1	Cover.
4	0.9	Ls, olive-gry-yel, wea yel-gry, f to med bioclas, v f xln mtrx, med bdd, fr resist, p expos.
3	10.1	Mostly cover; sh, varicolor, gry to prpl to red, wea prpl-gry, plty, sft, p expos, nonresist.
2	1.4	Ss, pnkish-red, wea dk ferrug-red, sb-qtzitc, v f grd, ang?, well srtd, med bdd, p resist and expos.
	199.3	Total thickness of Indian Springs formation; conformable contact with Battleship Wash formation.
		Battleship Wash formation
1	Not measured	Ls, lt gry, wea gry-bl to gry-tan, c bioclas, v f xln mtrx, mass, calc fracts.

SECTION H. ARROWLIME

Location.—NW ¼, NE ¼, NE ¼ sec. 27 to SW ¼, NW ¼, SE ¼ sec. 22, T. 18 S., R. 63 E., Clark County, Nevada.

Comments.—Section started at top of Battleship Wash formation along quarry exploration road in strike valley on west side of dolomitized Mississippian limestone ridge ½ mile northwest of Arrowlime Quarry (fig. 24). Line of section to northwest includes several offsets to avoid small faults and obtain better exposures. Exposures of both Indian Springs and Bird Spring formations poor to fair. Section measured primarily for thickness and facies control. UCLA locality 5263.

Unit	*Thickness*	*Lithologic description*
		Bird Spring formation
53	4.3	Ls, med gry, wea same, f bioclas, cptxln mtrx, blk ferrug chrty gossan on upper bdg surface, thn to med bdd, fr resist, rare *Chaetetes* heads up to 0.4′ in diam.
52	21.7	Alternating cover and ls, 45 percent expos; ls, lt gry, wea lt gry to tan, f to med bioclas, abdt dk brn chrt nods, cptxln mtrx, thn bdd, p expos.
51	6.3	Ls, lt to med gry, f to c bioclas, cptxln mtrx, med bdd, ptly silicif *Chaetetes* heads up to 2′ in diam, phaceloid rugose corals.
50	40.6	Cover except for a few thn bdd lt gry wea f bioclas nonresist ls, 10 percent expos.

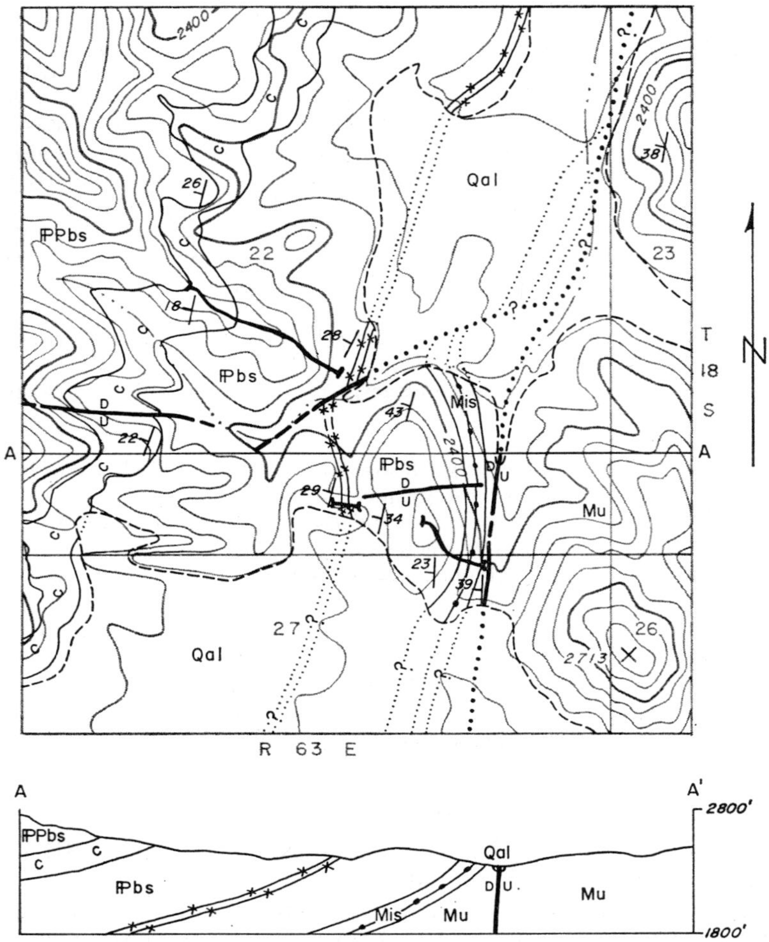

Fig. 24. Geologic map showing line of traverse for Section H, Arrowlime Quarry, Clark County, Nevada. See figure 21 for explanation of symbols.

Unit	*Thickness*	*Lithologic description*
49	11.2	Ls, lt gry, wea same, f to med bioclas, cptxln mtrx, sl aren, sml scale xbds, thn bdd, fr resist and expos.
48	22.4	Ls, lt gry, wea same, f bioclas, cptxln mtrx, v abdt dk brn chrt nods, resist, thn to med bdd, *Fusulinella* 6′ above base.
47	6.5	Ls, lt gry, wea same, f bioclas, cptxln mtrx, med to thck bdd, fr resist, dk brn chrt nods, fr abdt *Chaetetes* heads up to 2′ in diam.
46	80.5	Cover except for a 1′ thck ls 30′ above base and one 67′ above base; ls, lt gry, wea same, f to med bioclas, p expos.
45	9.6	Ls, lt to med gry, wea lt gry, f bioclas, cptxln mtrx, med bdd, dk brn wea chrt nods, s/ up to 1′ in diam, fr resist, rare *Chaetetes* heads up to 1′ in diam, in middle of unit.
44	143.2	Alternating cover and ls, 40 percent expos; ls, lt gry to med gry, wea lt gry to gry-tan, f to med bioclas, cptxln mtrx, aren in pt (f sb-ang well-srtd qtz), thn to med bdd, p resist and expos, few fos frag.

Unit	*Thickness*	*Lithologic description*
43	11.5	Ls, lt to med gry, wea lt gry to gry-tan, f bioclas, upper 8′ sl aren, lower 3′ v chrty, dk brn wea chrt nods, thn to med bdd, stylolitic, fr resist and expos.
42	15.0	Cover.
41	4.0	Ls, lt gry, wea tan, v aren, f bioclas, cptxln mtrx, dk brn chrt nods, thn to med bdd, p resist.
40	7.3	Ls lt gry, wea same, v f to med bioclas, cptxln mtrx, med to thck bdd.
39	37.9	Alternating cover and ls, 50 percent expos; ls, lt gry, wea lt gry and whish-gry, whish-gry where aren, grades into lmy ss at top, sd grs v f sb-ang well srtd, dk brn chrt nods, thn to med bdd, reworked by boring organisms.
38	101.5	Cover, float suggests interbdd chrty and nonchrty ls.
37	5.2	Sdy ls, lt gry, wea same, f bioclas, cptxln mtrx, med bdd, fr resist and expos, reworked, worm? borings.
36	42.4	Cover.
35	5.6	Ls, med gry, wea bl-gry, f bioclas, v aren, dk brn chrt nods and discont bnds, med bdd, resist.
34	26.0	Lmy sh, lt gry to prpl, wea cream to prpl, plty, sft, p expos.
33	18.3	Ls, med gry, wea bl-gry, f bioclas, aren in middle, dk brn chrt nods and discont bnds, cptxln mtrx, med bdd, fr resist and expos, *Paragassizocrinus* in upper 4′, *Osagia* in upper 2′.
32	32.7	Cover.
31	35.0	Ls, lt gry, wea rust-tan to brn, v f to f bioclas, cptxln mtrx, v aren, v f ang to sb-ang qtz, abdt dk brn chrt nods and discont bnds, thn bdd in a mass unit, fr resist and expos.
30	22.6	Ls, med gry, wea bl-gry, med to c bioclas, cptxln mtrx, scat dk brn chrt nods in lower 9′, med to thck bdd, resist, cliff former, abdt *Syringopora* 7′ above base and abdt scaphopods and *Bellerophon* 2′ above *S.*, few phaceloid corals scat throughout, unit intensely fractured.
29	6.9	Calc sh, lt gry to prpl, wea same, plty to bky, sft, nonresist, abdt brachs.
28	55.2	Ls, med gry, wea bl-gry, med to c bioclas, cptxln mtrx, lrg brn chrt nods, med to thck bdd, v resist, cliff former, *Bellerophon?* in upper half.
27	39.7	Ls, lt to med gry to blk, wea med bl-gry, f to c bioclas, dk brn chrt nods and discont bnds, v chrty, sl aren at top, thn bdd to thck bdd, resist, well exposed.
26	25.0	Cover.
25	23.3	Ls, lt to med gry to blk, wea med bl-gry, f to c bioclas, cptxln and arg mtrx, dk brn chrt nods and discont bnds, thck bdd, resist, good expos.
24	55.8	Ls, lt to med gry, wea lt bl-grn to dk gry, f to med bioclas, cptxln mtrx, abdt hash, arg in pt, dk brn chrt nods and some discont bnds, thn to med bdd, 40 percent expos, nonresist, slope former.
23	20.2	Ls, lt to med gry, wea lt gry, v f to med bioclas, cptxln mtrx, scat dk brn wea chrt nods and discont bnds, med to thck bdd, resist, cliff former, abdt *Orthotetes* at top.
22	2.4	Cgl, lt yel-brn, wea same to dk choc brn, pbbls up to ½″ diam, v aren, slt and sh mtrx, calc in pt, thn bdd, sft, nonresist, p to fr expos.
21	12.2	Interbdd ss, ls and sdy ls, lt gry, wea gry to tan, to rust-yel, f to

Unit	*Thickness*	*Lithologic description*
		med bioclas, sd grs f ang to sb-ang, cptxln mtrx, med to thck bdd, abdt brachs.
20	21.5	Ls, lt to med gry, wea gry, f to c bioclas, cptxln mtrx, dk brn wea chrt nods and discont bnds, med bdd, arg ptgs, abdt brachs.
	973.5	Total thickness of measured part of Bird Spring formation; conformable contact with Indian Springs formation.
		Indian Springs formation
19	5.2	Ls, lt gry, wea olive-gry to reddish-gry, f to c bioclas, resist, thn to thck bdd, abdt fos.
18	0.9	Ss, ferrug-brn, wea same, f to med grd, sb-ang to sb-rdd, p expos.
17	17.4	Cover.
16	4.1	Ls, gry to yel-gry, wea gry-yel, f to c bioclas, cptxln mtrx, med bdd, ferrug gossan on uppermost bddg surface w/ abdt spiriferids, fr resist and expos.
15	10.0	Cover.
14	3.9	Ls, gry-yel, wea yel, f to c bioclas, cptxln mtrx, thn to med bdd, fr resist and expos.
13	5.2	Cover.
12	1.0	Ls, yel-gry, wea ferrug-brn-yel, f to med bioclas, cptxln mtrx, med bdd, fr resist and expos.
11	0.6	Ss, ferrug-yel-orng, wea same, f grd, sb-rdd, well srtd, med bdd, p expos.
10	31.2	Cover.
9	1.1	Ls, dk gry, to rust-brn, wea rust-brn mottled w/ dk gry, f bioclas, cptxln mtrx, fr resist and expos, abdt fos.
8	9.7	Cover.
7	1.8	Ls, lt olive-gry, wea olive-red, f to c bioclas, v ferrug mtrx, upper 0.3′ dk red wea ferrug cgl, v p expos, ls fr resist and expos.
6	3.6	Qtzite, gry to tan, wea same, v f to f grd, dk brn ferrug gossan, fr resist and expos.
5	23.9	Cover.
4	3.6	Qtzite, lt olive-gry-yel, wea same or ferrug-brn where exposed for a period of time, v f to f grd, sb-ang to sb-rdd, plty, thn bdd, sml xbds, fr resist and expos.
3	6.5	Sh, varicolor, gry to yel to red, wea same, sl slty, bky to plty, sft, p expos.
2	21.0	Cover.
	150.7	Total thickness of Indian Springs formation; conformable contact with Battleship Wash formation.
		Battleship Wash formation
1	Not measured	Ls, lt to med gry, wea bl-gry or off-wh where dolomitized, f to c bioclas, cptxln mtrx, dnse, med to thck bdd, p expos.

SECTION I. APEX

Location.—SE ¼, NE ¼ to NW ¼, NW ¼, sec. 5, T. 19 S., R. 63 E., Clark County, Nevada.

Comments.—Section measured on gently west dipping east limb of broad north trending syncline (fig. 25). Indian Springs formation sharply folded and faulted and poorly exposed; Bird Spring formation fairly well exposed in the lower part, poorly exposed in the upper part and cut throughout by numerous small vertical faults of 2′ to 50′ displacement. UCLA locality 5274.

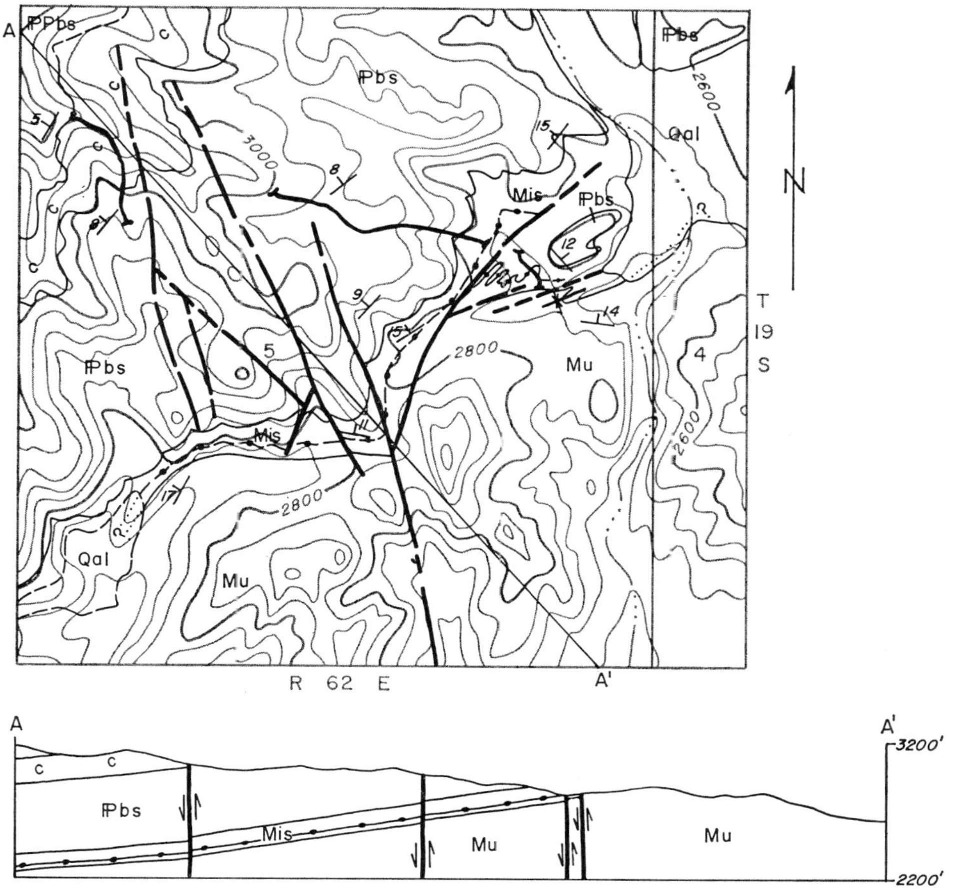

Fig. 25. Geologic map showing line of traverse for Section I, Apex, Clark County, Nevada. "Marker" limestone not present in this area. See figure 21 for explanation of symbols.

Unit	*Thickness*	*Lithologic description*
		Bird Spring formation
67	2.2	Ls, dk gry, wea med gry, v f to f bioclas, cptxln mtrx, dk brn chrt nods, med bdd, resist, *Chaetetes* heads up to 5″ diam in upper 1.0′.
66	14.4	Ls, lt gry, wea same, f to med bioclas, v f xln mtrx, thn to med bdd, abdt, dk brn chrt nods, p resist and 60 percent expos, abdt crin cols.
65	9.6	Ls, med gry, wea lt bl-gry, f to c bioclas, v f xln mtrx, dk brn chrt nods at top of unit, mass, ledge holder, sponge spicules? in upper 0.5′, silicif producs, *Chaetetes* heads up to 3.5′ long, others plty.
64	38.0	Cover 80 percent; ls, lt to dk gry, wea lt to med gry, thn bdd, v f to med bioclas, v f xln mtrx, p expos.
63	26.2	Ls, lt to med gry, wea lt gry, v f to med bioclas, sl aren, v f ang qtz, v f xln mtrx, thn to med bdd, scat dk brn chrt nods slope holder, lrg scale xbddg 5′ below top of unit.
62	5.7	Ls, dk gry, wea med bl-gry, f to med bioclas, v f xln mtrx, med bdd, cliff holder, *Fusulinella* in middle of unit.

Unit	*Thickness*	*Lithologic description*
61	12.5	Ls, lt gry, wea same, orng, chrt stringers, v f xln to cptxln, thn to med bdd, p resist, fr expos, slope holder.
60	4.9	Cover.
59	12.6	Ls, lt to med gry, wea lt gry, f to med bioclas, v f xln mtrx, thn to med bdd, p to fr resist and expos, *Fusulinella* in upper 0.5′.
58	8.4	Ls as in underlying exposed unit.
57	16.3	Cover.
56	4.7	Ls, lt gry, wea orng-tan, v f to cptxln, slty, chrty, knobby wea, p resist, fr expos.
55	8.7	Cover.
54	8.0	Ls, lt to med gry, wea bl-gry, f to med bioclas, cptxln to v f xln mtrx, dk brn chrt nods in upper 2′, med bdd, resist cliff former, s/ *Chaetetes* heads up to 1.0′ in diam in lower ½ of unit.
53	17.4	Ls, lt gry, wea gry-wh to lt gry-bl, f bioclas, v f xln mtrx, dk brn chrt nods, thn to med bdd, p resist, 50 to 60 percent expos.
52	6.0	Ls, lt gry-tan, wea lt gry, f to med bioclas, v f xln mtrx, mass, resist, ledge holder.
51	12.3	Ls, lt gry, wea same, f bioclas, v f xln mtrx, v chrty and aren (v f ang qtz), thn bdd, fr resist, slope holder, *Orthotetes* in lower pt.
50	15.8	Cover.
49	8.5	Ls, lt to med gry, wea lt gry, f to med bioclas, v f xln mtrx, aren (v f ang qtz), med to thck bdd, resist ledge former, sol corals in upper pt.
48	8.7	Cover.
47	24.7	Ls, lt gry, wea same, v f bioclas, cptxln mtrx, v aren in upper ½ (v f ang qtz), desert varnish where sdy, dk brn chrt nods in lower ⅓, thn to med bdd, upper 9′ cliff former, 70 percent expos, xbdd, in pt, sml scale festoons.
46	14.0	Ls, lt to med gry, wea gry-bl, v f to f bioclas, cptxln mtrx, stylolitic contact in upper pt, dk brn chrt nods up to 1.0′ in diam in upper bd, 60 percent expos, fr resist.
45	8.4	Ls, lt to dk gry, wea med bl-gry, v f to med bioclas, cptxln mtrx, dk brn chrty nods in lower 2.5′ w/ lrg echinoid spines and pls, sl aren, v f ang qtz, thn to med bdd, resist.
44	5.8	Cover.
43	26.7	Ls, lt gry, wea same, v f to med bioclas, v f xln mtrx, aren bnds grading into sdy ls in places, dk brn chrt in upper 5′, p to fr resist and expos, thn to med bdd, tan wea bnd at top of unit, yel wea bnd 14′ above base of unit.
42	10.3	Ls, lt gry, wea gry to tan, aren, v f ang qtz, dk brn chrt nods in upper pt, fr resist and expos, f to med bioclas, v f xln mtrx.
41	3.6	Cover.
40	19.4	50 percent expos: ls, lt gry, wea lt bl-gry, v f to med bioclas, v slty and sdy (v f ang qtz), v f xln mtrx, thn bdd, p resist and expos.
39	8.9	Ls, lt to med gry, wea lt gry w/ brn silicif nods and chrt nods, v f to f bioclas, v f xln mtrx, med bdd, resist, sdy in upper 1′.
38	7.5	Cover.
37	7.3	Ls, med gry, wea lt bl-gry, f bioclas, aren (v f ang qtz), v f xln mtrx, dk brn chrt nods, chrt bnd at top, med bdd, resist ledge holder.
36	20.3	Cover except for 2′ to 14′ above base of unit; ls, med gry, wea lt bl-gry, v f to f bioclas, slty, v f xln mtrx, dk brn chrt nods, thn bdd, p expos, silicif fos.
35	14.2	Ls, lt to med gry, wea lt bl-gry, v f to med bioclas, v f xln mtrx,

Unit	*Thickness*	*Lithologic description*
		abdt dk brn to blk chrt nods up to 2″ diam in lower ⅔, qtzitic bnds in upper ⅓ and *Osagia* in upper 2′, thn to med bdd, fr resist.
34	7.0	Ls, lt gry, wea lt bl-gry, ool, mass, resist, dk brn chrt nods up to 2′ in diam in middle of unit.
33	25.5	Cover.
32	8.8	Ls, med gry, wea lt bl-gry, v f bioclas, v f xln mtrx, chrty in lower ½, interbdd qtzite in upper ½, fr resist, thn to med bdd.
31	28.8	Arg ls, lt gry, wea same, slty, sft, punky on wea surface, nonresist, slope former, v chrty, blk and brn chrt nods, few sml producs, less than 10 percent expos.
30	3.4	*Syringopora* biostrome: ls, lt gry to med gry, wea lt bl-gry, v f xln, mass, p resist, phaceloid corals in base.
29	6.6	Ls, as in underlying unit w/ phaceloid corals at base.
28	55.8	Ls, med to dk gry, wea lt to med gry, v f to f xln, v f xln mtrx, bioclas, fauna as in underlying unit, abdt discont chrt bnds and nods, thn bdd in mass unit, may have a bench formed near middle of unit, resist, cliff former, abdt ramose bry and *Michelinia* in upper 10′.
27	41.0	Ls, dk gry to blk, wea lt gry, v f xln, arg, petro odor on fresh break, dk brn to blk chrt nods, thn bdd, 10 to 20 percent expos, mostly plty float, silicif brachs, slope former.
26	7.8	Ls, med to dk gry, wea lt bl-gry, v f bioclas, v f xln mtrx, dolomitic and chrt nods, mass w/ thn discont ptgs, silicif producs, cliff former.
25	29.0	Ls, lt to dk gry, wea med bl-gry, f to med bioclas, v f xln mtrx, thn to med bdd, approx 30 percent expos, slope holder, few scat dk brn to blk chrt nods, silicif brach frags.
24	6.6	Ls, dk gry-blk, wea lt bl-gry, sb-cptxln, dnse, blk to dk brn chrt nods scat through unit, cliff former.
23	10.8	Ls, med gry, f to med bioclas, wea med gry, v f xln mtrx, chrt pbbls, p resist, slope former, silicif brachs, med bdd.
22	17.6	Ls, med gry, wea same, v f to c bioclas, alternating v f and med to c zones, pbbl cgl at 7.4′ above base of unit and at top of unit, ls clasts in lower cgl, chrt in upper, chrty bnd in upper bd, good resist and expos, s/ silicif of brachs.
21	2.7	Sdy sh, lt gry-yel, wea same, bky, v f ang qtz grs, p resist, p expos.
20	1.0	Qtzite, yel-brn, wea same, v f to f grd, sml xbdd, thn bdd, p expos.
19	0.7	Cgl, lt to med gry, wea ferrug brn, ls pbbls, v f sd mtrx, ferrug, caps underlying ls w/ irreg contact, thckness variable, 0.2″ to 0.8″.
18	36.7	Ls, lt to med gry, wea med gry, med bioclas, v f xln mtrx, ferrug mtrx in pt, thn to med bdd, upper 8.0′ a good cliff former, lower pt p to fr expos.
	———	
	701.8	Total thickness of measured part of Bird Spring formation; conformable with Indian Springs formation.
		Indian Springs formation
17	11.3	Cover.
16	8.6	Ls, lt gry, wea rust-yel, med to c bioclas, ferrug and v f xln mtrx, med bdd, intensely fracted, fos not obtainable.
15	18.8	Mostly cover, float ls as in underlying unit, darker red, wea dk red and gry.
14	2.1	Ls, med gry-prpl, wea rust-orng-brn, med bioclas, v f xln and ferrug mtrx, med bdd, fr resist and expos, crin stems up to 8″ long.
13	21.8	Sh, blk to dk ferrug-red, wea red, spltry to plty to bky, nonresist, v p expos.

Unit	Thickness	Lithologic description
12	1.4	Ls, ferrug-brn, wea same, med to c xln, ferrug and v f xln mtrx, thn bdd, p resist.
11	11.0	Cover.
10	2.7	Ls, prpl-gry, wea ferrug-orng to tan, f to c xln, v f xln mtrx, ferrug mtrx in pt, thn to med bdd, p resist and expos.
9	18.4	Cover.
8	2.1	Ls, prpl-gry, wea ferrug-red-orng, f to med xln, v f xln mtrx, med bdd, lower 0.8′ cover.
7	3.0	Ls, med gry, wea ferrug-brn-red, f to c xln, v f xln mtrx, dk blk ferrug wea upper 0.5′ cgl, ramose bry, crin cols, med bdd.
6	1.4	Qtzite, pnk, v f grd, ferrug spots, med bdd.
5	9.4	Cover mostly, sh, blk, gry, red, wea ferrug-red, bky to plty, hard, brittle, slope former.
4	1.2	Ls, gry-brn, wea rust-brn, med xln, ferrug mtrx, thn bdd, p expos.
3	4.3	Qtzite, v f grd, siliceous mtrx, varicolor gry, prpl, orng, thn to med bdd, intensely fracted, sml xbdd, p expos.
2	14.5	Cover.
	132.0	Total thickness of Indian Springs formation; conformable contact with Battleship Wash formation.
		Battleship Wash formation
1	Not measured	Ls, lt bl-gry, wea lt tan-gry, f to med xln, v f xln mtrx, med to thck bdd, resist cliff former, authigenic qtz xls, no visible porosity.

SECTION J. GUNNERY RANGE

Location.—NW 1/4 sec. 8, T. 18 S., R. 62 E., Clark County, Nevada.

Comments.—Section measured on nearly vertical southeast limb of asymmetrical northeast plunging anticline (fig. 26). Indian Springs formation faulted and contorted; thickness of 268.8′ for this unit believed to be approximately 100′ excessive but cannot be proven because of poor exposures. Numerous small faults occur throughout Bird Spring formation and 100′ thick shear zone present in lower part of *Fusulinella* Zone. Inclusion of shear zone gives thickness of 910.9′ for Chester through Derry part of Bird Spring formation, which is thought to be approximately 100′ in excess at this locality. Section measured principally for stratigraphic thickness and facies information. UCLA locality 5276.

Unit	Thickness	Lithologic description
		Bird Spring formation
45	6.1	Ls, lt to med gry, wea bl-gry, f bioclas, cptxln mtrx, upper 3.4′ reworked w/ sol corals and *Syringopora*? thck bdd, *Chaetetes* heads up to 1′ in diam in lower 2.7′.
44	5.9	Sdy ls, lt gry, wea same, f bioclas, cptxln mtrx, med bdd, fr resist and expos.
43	6.2	Ls, lt to med gry, wea same, v f bioclas, cptxln mtrx, mass, resist, *Chaetetes* heads up to 1.3′ in diam.
42	35.0	Ls, med gry, wea lt gry, med to c bioclas, abdt dk brn to blk wea chrt nods and discont bnds, cptxln mtrx, cover in upper 11′, med bdd at base, thn bdd upward.
41	27.1	Ls, lt gry, wea prpl-gry, v f to med bioclas, arg, cptxln mtrx, plty, thn to med bdd, p resist, abdt producs.

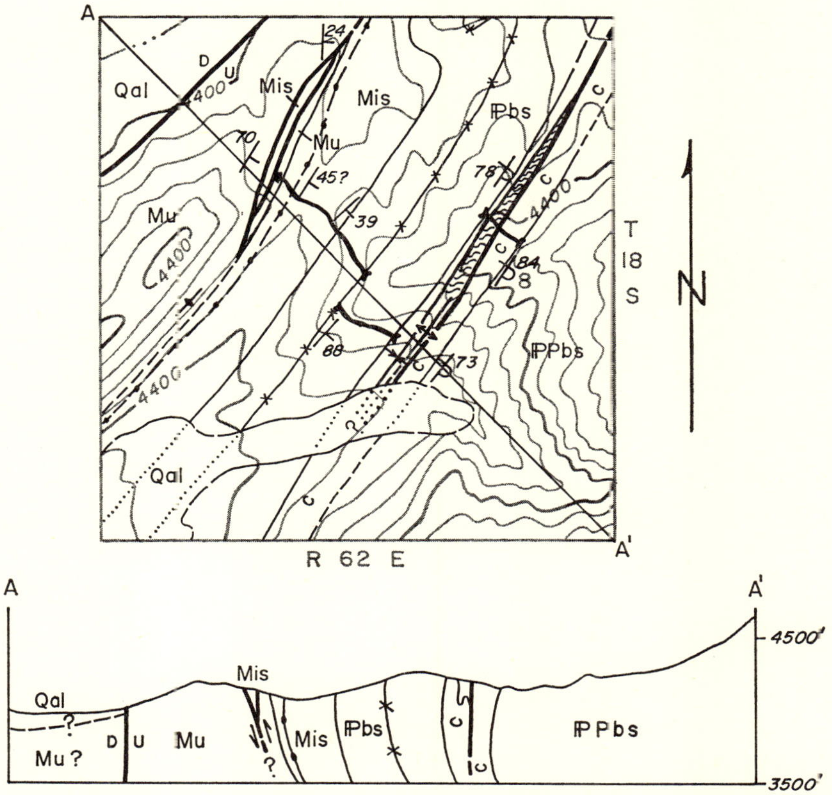

Fig. 26. Geologic map showing line of traverse for Section J, Gunnery Range, Clark County, Nevada. "Marker" limestone so thin it is expressed by a single line. See figure 21 for explanation of symbols.

Unit	*Thickness*	*Lithologic description*
40	7.0	Ls, lt gry-wh, wea rust-yel, slty, v f xln, cptxln mtrx, med bdd, p resist and expos.
39	38.4	Ls, lt to med gry, wea lt gry to whish in middle of unit, lt bl-gry elsewhere, f to med bioclas, cptxln mtrx, scat dk brn to blk wea chrt nods, med bdd, fr resist and expos, saddle former.
38	23.2	Ls, med gry, wea lt bl-gry, f to med bioclas, resist, thn to med bdd, *Fusulinella* 7′ to 10′ above base, dip 84°overturned, unit wea as a wall 6′ to 30′ high.
37	100.0	Shear zone, intensely brecciated, fractd, calcite filled fracts and veins.
36	11.1	Ls, lt to med gry, wea lt gry, med bioclas, cptxln mtrx, med bdd, aren (f sb-ang qtz) in lower pt, dk brn to blk wea chrt nods in base and top, fr resist and expos, intensely fract.
35	11.8	Cover.
34	6.1	Ls, lt to med gry, wea lt bl-gry, f bioclas, cptxln mtrx, med to thck bdd, fr resist and expos, *Chaetetes* heads up to 2.8′ diam and 1.5′ high, *Fusulinella* on top bddg surface.
33	28.4	Ls, lt gry, wea bl-gry, f to med bioclas, cptxln mtrx, thn to med bdd, scat dk brn wea chrt nods, fr resist and expos.

Unit	*Thickness*	*Lithologic description*
32	9.0	Ls grading upward into lmy ss, lt gry, wea tan, f bioclas, cptxln mtrx, sd grs f sb-ang well srtd, med bdd, fr resist and expos.
31	5.6	Cover.
30	188.2	Interval interbdd ls, sdy ls, chrty ls and sdy chrty ls, 75 percent expos: ls, lt to med gry, wea same, f to med bioclas, cptxln to v f xln mtrx, wea brn to tan where sdy, sd grns are v f to f sb-ang well srtd, thn to med bdd, fr resist, unit a ridge former, intensely fractd and numerous sml faults cut the interval, *Osagia* at base, two whish wea bnds, one 64′ above base, the other 12′ below top.
29	32.6	Ls, lt gry, wea lt gry-tan, v aren (v f to f sb-ang well srtd qtz), abdt brn wea chrt nods and discont bnds, thn bdd, plty in pt, wea in a mass unit.
28	16.4	Ls, med gry, wea bl-gry, f bioclas, cptxln mtrx, scat dk brn to blk chrt nods, med bdd, fr resist, 75 percent expos.
27	17.0	Cover.
26	5.8	Ls, lt to med bl, wea bl-gry, f to med bioclas, cptxln mtrx, chrt nods in lower ½, thck bdd upper ½ *Syringopora* biostrome.
25	15.0	Cover.
24	8.7	Ss grading upward into ls, lt gry, wea tan to gry, v f grd, sb-ang, med bdd, p resist and expos.
23	9.4	Ls, lt to med gry, wea med bl-gry, f to med bioclas, cptxln mtrx, scat chrt nods, most abdt in lower 2′, numerous calcite filled fracts, med to thck bdd, v resist, cliff former.
22	30.6	Cover.
21	37.3	Ls, lt gry, wea whish-gry, v f xln, cptxln mtrx, bky, scat dk brn to blk chrt nods, intensely fractd, would make a fr flagstone, thn to med bdd, fr resist, p expos.
20	46.2	Ls, med gry, wea lt to med gry, f bioclas, sl arg, s/ prpl shly bds in lower ½, abdt blk chrt nods, thn to med bdd, fr resist and expos.
19	15.2	Ls, med gry, wea lt gry, v f to f bioclas, arg, sl aren (v f sb-ang qtz), thn to mass bdd, resist.
18	61.1	20 percent expos, ls, med to dk gry, wea lt gry w/ prpl tint, dk brn to blk chrt nods, v f bioclas and arg, cptxln mtrx, abdt fos.
17	56.2	Cover, except for a thn blk plty sft fiss sh 18′ above base.
16	14.5	Ls, med gry, wea lt gry-tan w/ arg prpl interbds, s/ blk ls, wea lt gry, f bioclas, cptxln mtrx, thn to med bdd, (0.1′ to 2′), blk chrt nods and bnds.
15	7.4	Ls, lt to med gry, wea lt bl-gry, f to med bioclas, cptxln mtrx, s/ prpl pbbly arg ls, thn to med bdd, 0.2′ to 1.8′ silicif crin pls at base.
14	10.2	Ss at base grading upward into sdy ls, lt gry-tan, wea ferrug-brn at base and grades upward into lt gry, f grd, ang?, well srtd, xbdd, calc mtrx, thn to med bdd (0.4′ to 2.2′), resist, fr expos.
13	10.3	Cover.
12	7.9	Ls, lt to med gry, wea lt gry, med bioclas, cptxln mtrx, med bdd, calcite filled fracts, fr resist and expos.
	910.9	Total thickness of measured part of Bird Spring formation, approximately 100′ too thick owing to inclusion of shear zone; conformable contact with Indian Springs formation.
		Indian Springs formation
11	7.7	Cover.
10	42.2	Ls, lt gry, wea reddish-gry to yel-gry, f to med bioclas, cptxln mtrx, thn to med bdd, p resist and expos, 75 percent expos.
9	14.3	Cover.

Unit	*Thickness*	*Lithologic description*
8	7.3	Ls, lt gry-tan, wea red-brn, med bioclas, cptxln mtrx, med bdd, p resist and expos, cover in middle of unit.
7	147.3	Cover. Thickness probably 100′ excessive due to structural deformation in the interval.
6	2.2	Ls, lt gry, wea ferrug-red-brn, f to c bioclas, v f xln mtrx, med bdd, upper 0.6′ dk red-brn ferrug cgl, silicif and ls clasts up to 3.0′ diam, med bdd, p expos, conodonts.
5	3.8	Cover.
4	1.0	Qtzite, dk red-brn, ferrug mtrx in pt, dnse, hard, p expos.
3	40.5	Cover.
2	2.5	Ss, wh to tan, wea lt brn, f to med grd, calc mtrx, sb-ang, thn bdd, p expos.
	268.8	Total thickness of Indian Springs formation, approximately 100′ abnormally thick owing to deformation; conformable contact with Battleship Wash formation.
		Battleship Wash formation
1	Not measured	Ls, lt gry, wea bl-gry, med to c bioclas, v f xln mtrx, med to mass bbd, resist, crin cols, *Michelinia?*, sol rugose corals.

SECTION K. GASS PEAK

Location.—NE ¼ NE ¼ sec. 22, and N ½, SW ¼, NW ¼ sec. 23, T. 18 S., R. 61 E., Clark County, Nevada.

Comments.—Section measured on north dipping limb of eastward plunging syncline (fig. 27). Exposures of Indian Springs and Bird Spring formations fair to good. Numerous offsets in line of section required because of faulting and minor folding, particularly in Indian Springs formation. UCLA locality 5277.

Unit	*Thickness*	*Lithologic description*
		Bird Spring formation
69	16.5	Ls, lt to med gry, wea lt gry, f to c bioclas, v f xln mtrx, wavy thck bdd, prominent cliff former, aren bd in middle of unit, separates two *Chaetetes* bds, *C.* 4′ above base and 4′ below top, upper 3.8′ contain sol rugose corals, numerous *C.* heads cylindrical shaped.
68	7.5	Cover.
67	16.3	Ls, lt to med gry, wea lt gry-bl, med bioclas, v f xln mtrx, blk chrt nods in upper ½, thn to thck bdd, cliff former, sol corals in upper 1′.
66	10.0	Cover.
65	12.4	Ls, lt gry, wea yel-gry, med bioclas, arg ptgs, v f xln mtrx, nonresist, thn bdd.
64	20.2	Cover.
63	9.6	Ls, lt to med gry, wea lt gry w/ prpl cast, f to c bioclas, cptxln to v f xln mtrx, arg in pt, p expos.
62	32.6	Cover.
61	1.5	Ls, lt gry, wea same, bioclas, v f xln mtrx, chrt nods, thn bdd, 0.6′.
60	14.0	Cover.
59	33.2	Ls, med gry, wea lt bl-gry, f to c bioclas, v f xln mtrx, dk gry chrt nods up to 1′ in diam, thn to thck bdd, 0.1′ to 5′, arg just below middle, intensely fractd, *Fusulinella* and *Syringopora* in lower 10′, large echinoid spines, sol corals, brach frags in upper 5′, cliff former.
58	21.2	Cover.
57	11.4	Ls, lt gry, wea lt gry-yel, f to c bioclas, arg in lower ½, v f xln to

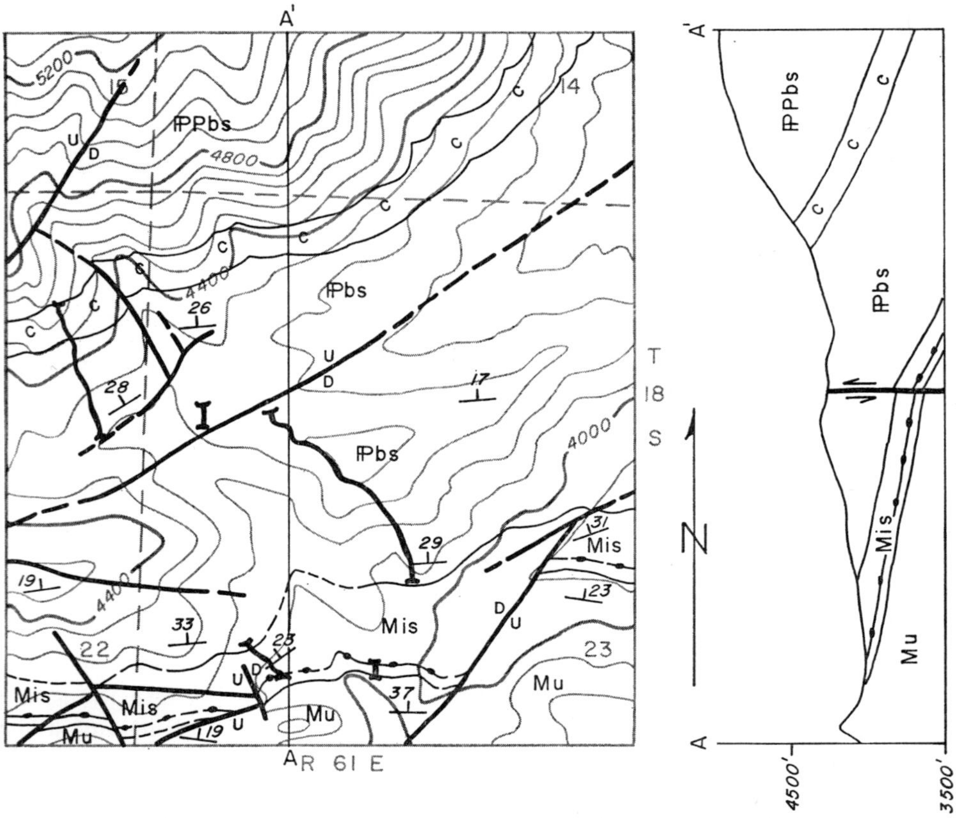

Fig. 27. Geologic map showing line of traverse for Section K, Gass Peak, Clark County, Nevada. "Marker" limestone not present in this area. See figure 21 for explanation of symbols.

Unit	*Thickness*	*Lithologic description*
		cptxln mtrx, wavy bdd, dk gry to blk chrt nods, p resist and expos, *Chaetetes* heads up to 1.5′ long, 1′ in diam.
56	10.6	Cover.
55	21.9	Ls, med gry, wea lt gry, f to c bioclas, v f xln mtrx, chrt nods wea w/ ferrug gossan, aren at base, arg in middle, fr resist, good expos.
54	28.2	Cover except for lt gry wea v f to c bioclas aren 2′ thck ls 12′ above base.
53	7.7	Ls, lt to med gry, wea lt gry, med bioclas, v f xln mtrx, upper ½ ool, scat dolo nods in lower ½, thck bdd, resist
52	21.8	Ls, lt gry, wea same w/ ferrug brn bnds, f to c bioclas, v f xln mtrx, abdt brn stnd dk gry dolo and chrt nods, thn to med bdd (0.6″ to 1.5′), cliff former.
51	61.5	Alternating cover and p expos ls, 45 percent expos: ls, lt to dk gry, wea med bl-gry, w/ aren bds wea lt gry-tan, f to c bioclas, v f xln mtrx, dk gry chrt nods, thn to med bdd (0.2′ to 1.5′), s/ pit and cusp wea, aren bds, xbdd, worm? trails in lower pt, *Rhipidomella elyensis, Orthotetes,* and sol corals in lower pt, phaceloid corals in upper pt.
50	12.2	Ls, lt to med gry, wea lt gry-tan, f to med bioclas, v f xln mtrx, abdt, chrt nods, sl aren, bddg discont, 0.1′ to 2′ thck, surfaces sb-plnr, resist cliff former, *Osagia?* in upper 1′.

Unit	*Thickness*	*Lithologic description*
49	14.1	Ls, med to dk gry, wea bl-gry, f to c bioclas, v f xln mtrx, lower 5′ sl aren (v f grd ang qtz), fr even bddg surfaces, bddg 0.3′ to 3′ thck, scat dk gry chrt nods, s/ pit and cusp wea, cliff former, silicif *Orthotetes?*.
48	34.6	Ls, dk gry, wea lt bl-gry, v f to c bioclas, cptxln mtrx, dk gry to blk chrt nods in discont bnds, thn to med bdd, 30 percent cover, fr resist, p expos.
47	26.0	Ls, lt to med gry, wea lt gry-tan, f to c bioclas, cptxln mtrx, v aren slt and v f grd ang qtz, dk gry chrt nods in lower pt, dolo nods in upper ½, ferrug stng around nods, cover 60 percent.
46	33.6	Ls, lt to med gry, wea med gry-bl alternating w/ brn-rust bnds, f to c xln, cptxln mtrx, abdt dk gry-brn wea dolo nods in s/ bds, bddg persistent (0.4′ to 6′), ledge former, few obtainable fos, lrg (1″ diam) ramose bry at base, scat silicif brachs.
45	6.3	Ls, lt gry, wea cream-yel, v f xln, cptxln mtrx, aren (v f grd), numerous ferrug coated dolo and chrt nods, thn bdd in mass unit, prominent color stripe on hillside.
44	16.8	Cover.
43	14.2	Ls, lt to med gry, wea bl-gry w/ rust-brn bnds, med bioclas, v f xln mtrx, chrt and dolo nodular bnds, nods wea w/ ferrug gossan, even bddg (0.6′ to 2.2′), resist, cliff former, *Osagia* in upper 10′.
42	9.2	Cover.
41	3.8	Ls, lt to med gry, wea lt gry w/ rust splotches, med bioclas, v f xln mtrx, thn to med bdd, fr resist, dk brn chrt nod in discont bnds.
40	11.4	Cover.
39	23.4	Ls, lt to med gry, wea lt gry w/ rust splotches, med bioclas, v f xln mtrx, c bioclas upward, chrt and dolo nods as in underlying unit, thn to med bdd in lower ½, med to thck bdd in upper ½, resist, cliff former, few silicif brachs.
38	6.5	Cover.
37	13.7	Ls, med gry, wea bl-gry, med bioclas, v crinoidal, cptxln to v f xln mtrx, nod chrt and dolo in discont bnds, xbdd (0.5″ to 0.7″ long, 0.5″ to 2″ deep), even bddg 1″ to 1.5′, cliff former.
36	22.6	Cover.
35	31.1	Ls, lt to med gry, wea lt gry w/ brn ferrug bnds, v f bioclas, cptxln mtrx, abdt ferrug brn chrt and dolo nods in discont bnds, thn bdd (0.1′ to 1′), few silicif brachs.
34	13.3	Ls, med gry, wea bl-gry med bioclas, v f xln mtrx, dk brn to blk chrt nods in discont bnds, s/ dolo nods, ferrug gossans on nods med bdd (0.2′ to 2′), surfaces sb-plnr, resist cliff former, fos most abdt 3.5′ above base of unit.
33	63.2	30 percent expos: where exposed ls, lt to med gry, wea gry-tan, med bioclas, cptxln to v f xln mtrx, abdt dk brn to blk chrt nods in discont bnds, s/ dolo nods, ferrug gossans on nods, thn bdd (0.2′ to 1′), surfaces sb-plnr to undulty, aren in pt, few fos 42′ above base of unit, silicif brachs.
32	72.0	Ls, lt to dk gry, wea lt gry lower 40′ and med gry in upper 32′, med bioclas, v f xln mtrx, chrt and dolo nods in discont bnds w/ ferrug gossans, dk brn to blk interiors, thn to med bdd in lower pt, thcker bdd upward, few producs and other brachs in lower pt, upper pt abdt fos, bry and corals abdt 41′ above base, calc filled fracts, good resist, prominent cliff former
31	36.8	Cover.

Unit	*Thickness*	*Lithologic description*
30	12.6	Ls, med to dk gry, wea med gry, v f bioclas, cptxln mtrx, aren in middle and top of unit, both aren bds wea lavender, chrt nods wea w/ ferrug gossan, med bdd, resist, fos in upper half.
29	31.5	Ss, lt gry, wea orng-brn to rust, v calc, v f to f grd, ang, well srtd, xbdd, thn to thck bdd, 0.1′ to 1.5′, interbdd ls in upper 10′, ls lt to med gry, wea lt gry, med bioclas, v f xln mtrx, sl aren, 0.3′ to 1′ bddg, few fos.
28	63.6	Ls, lt to dk gry, wea lt to med gry, med bioclas, v f xln mtrx, arg interbds, nod dk brn to blk chrt bnds discont, aren bnds in upper 6′, thn to thck bdd (0.5″ to 2′), 50 percent expos, slope former, abdt fos, most fos found in float, upper 10′ resist ledge former.
27	66.6	Ls, lt to med gry, wea lt gry to prpl, f bioclas, arg, cptxln mtrx, nod dk brn to blk chrt in discont bnds, upper 12.6′ more chrty and well expos, remainder 40 to 50 percent expos, thn bdd, v fos, most fos collected from float and may contain a few from overlying bds, 2 thn (2″) cgl bds found in float pieces.
26	52.0	Cover, 1 thn 0.6′ ls, lt gry, wea yel, f bioclas, sft, mollusk fauna, at base.
25	4.9	Ls, med gry to prpl-gry, wea same, med to c bioclas at base becoming finer upward, cptxln mtrx, petro odor on fresh break, sb-plnr bddg 0.5′ to 1′ thck, p expos.
24	5.0	Ls, lt to med gry, wea lt gry, f to c bioclas, cptxln mtrx, even bdd, 3″ to 1.5′, resist, silicif fos, abdt crin ossicles in upper 3′.
23	6.4	Ls, lt to med gry, wea lt gry, f to med bioclas, cptxln mtrx, scat blk to dk brn chrt nods, aren ptgs, med bdd, resist, silicif brachs.
22	5.4	Ss, lt gry, wea lt brn, calc cmt, v f grd, sb-ang?, well srtd, middle 0.6′ sdy gry ls, upper bd 2.5′ xbdd, lower bd 2.3′ uniform even bdd ss which looks like ls on fresh break, resist.
21	19.8	Ls, lt gry, wea same, bioclas, lower 3.3′ shly, med bioclas, calc filled fracts, sb-plnr bddg 2″ to 2′, thck bdd upward, abdt fos, silicif in pt.
	———	
	1,070.9	Total thickness of measured part of Bird Spring formation; conformable contact with Indian Springs formation.
		Indian Springs formation
20	7.8	Ls, lt to med gry, wea lt gry w/ yel cast, med bioclas, cptxln mtrx, sb-plnr bddg 0.5′ to 2′ thck, spiriferids (particularly abdt on tops of bds), *Archimedes* in lowermost bd, fr resist and expos.
19	7.4	Cover.
18	5.8	Ls, lt to med gry, wea lt yel-gry, f to c bioclas, cptxln mtrx, bddg 0.5′ to 1′, evenly bdd, fr resist and expos.
17	12.2	Cover.
16	14.2	Ls, lt gry to lt brn, wea lt gry to rust-brn-red, bioclas, thn to med bdd (1″ to 1.5′), sb-plnr surfaces, resist, fr expos.
15	51.9	Cover, *Michelinia* in this unit.
14	1.9	Ls, dk rust-brn, wea orng-brn, f to med bioclas, cptxln mtrx, ferrug mtrx in pt, irreg uneven bddg.
13	2.6	Sh, dk gry to gry-grn, wea gry-grn, plty, nonresist, p expos.
12	11.7	Cover, probably interbdd sh and ls.
11	6.0	Ls, lt gry, wea ferrug-yel to red, f to c bioclas, cptxln calc and ferrug mtrx, bddg 0.5″ to 1′ thck, brachs; upper 0.7′ dk red to brn ferrug cgl, ls and silicif clasts up to 3″ in diam.
10	9.7	Cover.

Unit	*Thickness*	*Lithologic description*
9	6.4	Ls, lt gry to brn, wea rust-yel to reddish-brn, f to c bioclas, cptxln mtrx, bddg 4″ to 1.5′, fr resist.
8	4.3	Cover.
7	9.1	Ss, gry to lavender, wea same, v f grd, qtzite, sb-ang?, laminated to thn bdd, resist, sml festoon xbds (1″ to 8″).
6	1.7	Ls, lt gry to rust, wea rust-yel, f to c bioclas, cptxln mtrx, thn to med bdd (4″ to 1′), p resist.
5	4.6	Ss, lt gry, wea dk rust-brn, v f grd, sb-ang?, qtzitic in pt, silicious cmt, sml festoon bnds (1″ long, ⅛″ to ¼″ deep), laminated to thn bdd, grades upward into sdy ls to ls.
4	0.9	Ls, lt gry, wea rust-yel, f to med bioclas, cptxln mtrx, thn bdd, p expos, abdt brachs.
3	19.6	Cover.
2	1.4	Ss, lt brn, wea lt prpl-gry to tan, v f grd, sb-ang, qtzitic in pt, sml scale xbds (1″ long, ⅛″ deep), thn bdd (1″ to 4″), p expos.
	179.2	Total thickness of Indian Springs formation; conformable contact with Battleship Wash formation.
		Battleship Wash formation
1	Not measured	Ls, gry-bl, wea same, f to med bioclas, cptxln mtrx, thn to thck bdd, prominent ridge former.

LITERATURE CITED

BECKMANN, H.

1949. Conodonten aus dem Iberger Kalk (Ober-Devon) des Bergischen Landes und ihr Reinban. Senckenbergiana, 30:153–168, pls. 1–4.

BISCHOFF, G.

1957. Die Conodonten-Stratigraphie des rheno-herzynischen Unterkarbons mit Berücksichtigung der Wocklumeria-Stufe und der Devon/Karbon-Grenze. Hess. Landesamt Bodenforsch, Abhundl., Heft 19. 64 pp., 6 pls., 1 fig.

BISCHOFF, G., and W. ZIEGLER

1956. Das Alter der "Urfer Schichten" im Harburger Hinterland nach Conodonten. Hess. Landesamt Bodenforsch, Notiz, 84:138–169, pls. 11–14.

BISSELL, H. J.

1962. Pennsylvanian and Permian rocks of Cordilleran area. *In* C. C. Branson, ed., Pennsylvanian System in the United States. Amer. Assoc. Pet. Geol. Pp. 188–263, 31 figs.

BOUCKAERT, J., and A. C. HIGGINS

1963. La Base du Namurien dans le Bassin de Dinant. Bull. Soc. Belg. Geol., Paleontol., & Hydrol., T. LXXII, fasc. 2. 17 pp., 7 pls., 3 figs.

BOWYER, B. E., E. G. PAMPEYAN, and C. R. LONGWELL

1958. Geologic map of Clark County, Nevada. U.S. Geol. Surv. Mineral Invest. Field Stud. Map MF-138.

BRANSON, E. B., and M. G. MEHL

1933–1934. Conodont Studies. Univ. Missouri Stud., vol. 8. 349 pp., 28 pls.

1941*a*. Caney conodonts of Upper Mississippian age. Denison Univ. Bull., J. Sci. Lab., XXXV: 167–178, pl. v.

1941*b*. New and little known Carboniferous conodont genera. J. Paleontol., 15:97–106, pl. 19.

1944. Conodonts. *In* Shimer, G. W., and R. R. Shrock, Index Fossils of North America. New York: John Wiley. Pp. 225–246.

BRILL, K. G.

1963. Permo-Pennsylvanian stratigraphy of western Colorado Plateau and eastern Great Basin Regions. Geol. Soc. Amer. Bull., 74:307–331, 1 pl., 17 figs.

CASSITY, P. E., and R. L. LANGENHEIM, JR.

1966. Pennsylvanian and Permian fusulinids of the Bird Spring Group from Arrow Canyon, Clark County, Nevada. J. Paleontol., 40:931–968, pls. 110–114.

CLARKE, W. J.

1960. Scottish Carboniferous conodonts. Trans. Edinburgh Geo. Soc., 18:1–31, pls. i–v, 1 table.

COLLINSON, C., A. J. SCOTT, and C. B. REXROAD

1962. Six charts showing biostratigraphic zones, and correlations based on conodonts from the Devonian and Mississippian rocks of the Upper Mississippi Valley. Illinois Geol. Surv. Circ. 328. 32 pp.

COOGAN, A. H.,

1964. Early Pennsylvanian history of Ely Basin, Nevada. Amer. Assoc. Pet. Geol. Bull., 48: 487–495, 6 figs.

COOPER, C. L.

1939. Conodonts from a Bushberg-Hannibal Horizon in Oklahoma. J. Paleontol., 13:379–422, pls. 39–47, 2 text figs.

1947. Upper Kinkaid (Mississippian) microfauna from Johnson County, Illinois. J. Paleontol., 21:81–94, pls. 19–23, 1 text fig.

DOTT, R. H., JR.

1955. Pennsylvanian stratigraphy of Elko and northern Diamond ranges, northeastern Nevada. Amer. Assoc. Pet. Geol. Bull., 39:2211–2305.

DUNN, D. L.

1965. Late Mississippian conodonts from the Bird Spring formation in Nevada. J. Paleontol., 39:1145–1150, pl. 140, 2 figs.

1966. New Pennsylvanian platform conodonts from southwestern United States. J. Paleontol., 40:1294–1303, pls. 157, 158.

EASTON, W. H., and others
1953. Revision of stratigraphic units in Great Basin. Amer. Assoc. Pet. Geol. Bull., 37:143–151, 2 figs.

ELIAS, M. K.
1956. Upper Mississippian and Lower Pennsylvanian formations of south-central Oklahoma. *In* Petroleum geology of southern Oklahoma. Ardmore Geol. Soc., 1:56–134, pls. i–vi.
1959. Some Mississippian conodonts from the Ouachita Mountains. *In* The Geology of the Ouachita Mountains: A symposium. Dallas and Ardmore Geol. Soc. Pp. 141–165, 2 pls.
1966. Late Paleozoic conodonts from the Ouachita and Arbuckle Mountains of Oklahoma. Okla. Geol. Surv., Guide Book XVI. 39 pp., 2 pls.

ELLISON, S.
1941. Revision of the Pennsylvanian conodonts. J. Paleontol., 15:107–143, pls. 20–23, 4 text figs.
1944. *In* Branson, E. B., The Geology of Missouri. Univ. Missouri Stud., Vol. 19, no. 3. 535 pp., 49 pls.

ELLISON, S., and R. W. GRAVES, JR.
1941. Lower Pennsylvanian (Dimple Limestone) conodonts of the Marathon Region, Texas. Univ. Missouri School Mines and Metallurgy Bull., vol. 14, no. 3. 19 pp., 3 pls., 1 text fig.

FLÜGEL, H., and W. ZIEGLER
1957. Die Gliederung des Oberdevons and Unterkarbons am Steinbergwestlich von Graz mit Conodonten. Naturwiss. Vereines für Steiermark, Mitteilungen, 87:25–60, pls. i–v, 6 figs.

GLOBENSKY, Y.
1967. Middle and Upper Mississippian conodonts from the Windsor Group of the Atlantic Provinces of Canada. J. Paleontol., 41:432–448, pls. 55–58.

GORDON, M., JR.
1965. Carboniferous cephalopods of Arkansas. U.S. Geol. Surv. Prof. Paper 460. 314 pp., 96 figs., 32 pls.

GUNNELL, F. H.
1931. Conodonts from the Fort Scott Limestone of Missouri. J. Paleontol., 5:244–252, pl. 29.
1933. Conodonts and fish remains from the Cherokee, Kansas City, and Wabaunsee groups of Missouri and Kansas. J. Paleontol., 7:261–297, pls. 31–33.

HARLTON, B. H.
1933. Micropaleontology of the Pennsylvanian Johns Valley shale of the Ouachita Mountains, Oklahoma, and its relationship to the Mississippian Caney shale. J. Paleontol., 7:3–29, pls. 1–7.

HARRIS, R. W., and R. V. HOLLINGSWORTH
1933. New Pennsylvanian conodonts from Oklahoma. Amer. J. Sci., 225:193–204, 1 pl.

HASS, W. H.
1941. Morphology of conodonts. J. Paleontol., 15:71–81, pls. 12–16.
1951. Age of Arkansas Novaculite. Amer. Assoc. Pet. Geol. Bull., 35:2526–2541, 1 pl.
1953. Conodonts of the Barnett formation of Texas. U.S. Geol. Surv. Prof. Paper 243–F. Pp. 69–94, pls. 14–16.
1962. Conodonts. *In* Treatis on Invertebrate Paleontology, R. C. Moore, ed. (w) Miscellanea. Pp. 3–69, 42 figs.

HEATH, C. P., D. N. LUMSDEN, and A. V. CAROZZI
1967. Petrography of a carbonate transgressive-regressive sequence: the Bird Spring group (Pennsylvanian) Arrow Canyon Range, Clark County, Nevada. J. Sed. Petrol., 37:377–400, 10 figs.

HEWETT, O. F.
1931. Geology and ore deposits of the Goodsprings Quadrangle, Nevada. U.S. Geol. Surv. Prof. Paper 162. 172 pp., 55 figs., 40 pls.

HIGGINS, A. C.
1961. Some Namurian conodonts from North Staffordshire: Geol. Mag., 98:210–224, pls. x–xii, 6 text figs.
1962. Conodonts from the "Griotte" Limestone of North-West Spain. Inst. Geol. y Minero de Espana, Notes y Commun., 65:5–21, text figs. 1–2, 3 pls.
HINDE, G. J.
1900. Notes and descriptions of new species of Scotch Carboniferous conodonts. Trans. Natur. Hist. Soc. Glasgow, n.s., 5:338–346, pls. 9, 10.
HOLMES, G. B.
1928. A bibliography of the conodonts with descriptions of early Mississippian species. U.S. Natur. Mus. Proc., vol. 72, art. 5, no. 2701. 38 pp., 11 pls.
IGO, H., and T. KOIKE
1964. Carboniferous conodonts from the Omi Limestone, Niigata Prefecture, Central Japan. Studies of Asian conodonts, part I. Trans. Proc. Palaeontol. Soc. Japan, n.s., 53:179–193, pls. 27, 28.
1965. Carboniferous conodonts from Yobara, Akiyoshi Limestone, Japan. Studies of Asiatic conodonts, Part II. Trans. Proc. Palaeontol. Soc. Japan, n.s., 59:83–91, pls. 8, 9.
JOHNSON, M. S., and D. E. HIBBARD
1957. Geology of the Atomic Energy Commission Nevada Proving Grounds Area, Nevada. U.S. Geol. Surv. Bull. 1021–K. Pp. 333–384.
KOIKE, T.
1967. A Carboniferous succession of conodont faunas from the Atetsu Limestone in southwest Japan. Sci. Rep. Tokyo Kyoiku Daigaku Sect. C, no. 93, 9:279–318, 4 pls.
LANE, H. R.
1967. Conodonts from the uppermost Mississippian and Lower Pennsylvanian of the type Morrowan region, Arkansas. J. Paleontol., 41:920–942, pls. 119–123.
LANE, N. G.
1961. An occurrence of Archimedes in southern Nevada. J. Paleontol., 35:874–875, 1 fig.
1963. A silicified Morrowan brachiopod faunule from the Bird Spring formation, southern Nevada. J. Paleontol., 37:379–392, pls. 43–45, 6 figs.
1964*a*. New Pennsylvanian crinoids from Clark County, Nevada. J. Paleontol., 38:677–684, pl. 112, 1 fig.
1964*b*. Costation patterns in early Pennsylvanian spiriferids. J. Paleontol., 38:781–785.
LANE, N. G., and G. D. WEBSTER
1966. New Permian crinoids from southern Nevada. Univ. Calif. Pub. in Geol. Sci., vol. 63. Univ. Calif. Press. 87 pp., 13 pls.
LANGENHEIM, R. L., JR.
1963*a*. Mississippian stratigraphy in southern Utah and adjacent parts of Nevada and Arizona. *In* Geology of southwestern Utah. Intermountain Assoc. Pet. Geol. 14th Ann. Field Conf. Guidebook. Pp. 30–41.
1963*b*. Nomenclature of the Late Mississippian White Pine shale and associated rocks in Nevada. Illinois State Acad. Sci. Trans., 55:133–145, 3 figs.
LANGENHEIM, R. L., JR., and others
1962. Paleozoic section in Arrow Canyon range, Clark County, Nevada. Amer. Assoc. Pet. Geol. Bull., 46:592–609, 5 figs.
LANGENHEIM, V. A. M., and R. L. LANGENHEIM, JR.
1965. The Bird Spring group, Chesterian through Wolfcampian, at Arrow Canyon, Arrow Canyon range, Clark County, Nevada. Trans. Illinois Acad. Sci., 58:225–240, 6 figs.
LINDSTRÖM, M.
1964. Conodonts. Amsterdam: Elsevier. 196 pp., 64 figs.
LONGWELL, C. R.
1921. Geology of the Muddy Mountains, Nevada, with a section to the Grand Wash Cliffs in western Arizona. Amer. J. Sci., 201:39–62, 5 figs.
1928. Geology of the Muddy Mountains, Nevada. U.S. Geol. Surv. Bull. 798. 152 pp., 17 pls., 9 figs.

LONGWELL, C. R., and C. O. DUNBAR
1936. Problems of Pennsylvanian-Permian boundary in southern Nevada. Amer. Assoc. Pet. Geol. Bull., 20:1198–1207, 6 figs.
LONGWELL, C. R., E. H. PAMPEYAN, B. BOWYER, and R. J. ROBERTS
1965. Geology and mineral deposits of Clark County, Nevada. Nev. Bur. Mines Bull. 62. 218 pp., 16 pls., 22 figs.
LYS, M., and B. SERRE
1957. Presence de conodontes dans le Paleozoique du Sahara (region d'Adrar-Tarizrouft). Compt. Rend. Acad. Sci. Paris, 244:916–918.
1958. Contribution a la Connaissance des Microfaunes du Paleozoique. Revue Inst. tut Francais Petrole, 13:879–916, pls. i–x, tables 1, 2.
MCCUTCHEON, V. A., and E. C. WILSON
1961. *Ptolemaia,* a new colonial rugose coral from the Lower Permian of eastern Nevada and western Russia J. Paleontol., 35:1020–1028, pl. 123, 3 figs.
MCLAUGHLIN, K. P.
1952. Microfauna of the Pennsylvanian Glen Eyrie formation, Colorado. J. Paleontol., 26:613–621, pls. 82, 83.
MACURDA, D. B., JR.
1964. The blastoid *Pentremites* Say: A trimerous mutant and some new occurrences. J. Paleontol., 38:705–710, pl. 116, 2 figs.
MEISCHNER, K. D.
1962. Rhenaer Kalk und Posidonien Kalk im Kulm des nordostlichen Rheinischen Schiefergebirges und der Kohlenkalk von Schreufa (Eder). Hess. Landesamt Bodenforsch., Abhandl, Heft 39. 47 pp., 6 pls.
MOORE, R. C., and others
1944. Correlation of Pennsylvanian formations of North America. Geol. Soc. Amer. Bull., 55: 657–706, 1 chart.
MURRAY, F. N., and J. CHRONIC
1965. Pennsylvanian conodonts and other fossils from insoluble residues of the Minturn formation (Desmoinesian), Colorado. J. Paleontol., 39:594–610, pls. 71–73.
NOLAN, T. B., and others
1956. The stratigraphic section in the vicinity of Eureka, Nevada. U.S. Geol. Surv. Prof. Paper 276. 77 pp., 2 pls.
REXROAD, C. B.
1957. Conodonts from the Chester series in the type area of southwestern Illinois. Illinois Geol. Surv., Rep. Invest. 199. 43 pp., 4 pls, 21 text figs.
1958. Conodonts from the Glen Dean formation (Chester) of the Illinois Basin. Illinois Geol. Surv. Rep. Invest. 209. 27 pp., 6 pls., 14 text figs.
REXROAD, C. B., and R. C. BURTON
1961. Conodonts from the Kinkaid formation (Chester) in Illinois. J. Paleontol., 35:1143–1158, pls. 138–141, 4 text figs.
REXROAD, C. B., and C. COLLINSON
1961. Preliminary range chart of conodonts from the Chester series (Mississippian) in the Illinois Basin. Illinois Geol. Surv., Circ. 319. 11 pp., 1 chart.
1963. Conodonts from the St. Louis formation (Valmeyeran series) of Illinois, Indiana, and Missouri. Illinois Geol. Surv., Circ. 355. 28 pp., 2 pls., 5 text figs.
REXROAD, C. B. and W. M. FURNISH
1964. Conodonts from the Pella formation (Mississippian), south-central Iowa. J. Paleontol., 38:667–676, pl. 111.
REXROAD, C. B., and H. R LANE
1966. Classification of *Cavusgnathus alta,* type species of the conodont genus *Cavusgnathus.* J. Paleontol., 40:1391–1392, 1 fig.
REXROAD, C. B., and R. M. LIEBE
1962. Conodonts from the Paoli and equivalent formations in the Illinois Basin. Micropaleontol., 8(4):509–514.

RHODES, F. H. T.
1952. A classification of Pennsylvanian conodont assemblages. J. Paleontol., 26:886–901, pls. 126–129, 4 text figs.

RICH, M.
1961. Stratigraphic section and fusulinids of the Bird Spring formation near Lee Canyon, Clark County, Nevada. J. Paleontol., 35:1159–1180, pls. 142–146.
1963. Petrographic analysis of Bird Spring group (Carboniferouspermian) near Lee Canyon, Clark County, Nevada. Amer. Assoc. Pet. Geol. Bull., 47:1657–1681.

ROUNDY, P. V., G. H. GIRTY, and M. I. GOLDMAN
1926. Mississippian formations of San Saba County, Texas. U.S. Geol. Surv. Prof. Paper 146. 63 pp., 33 pls. Microfauna by Roundy, 5–23, pls. i–iv.

SADLICK, W.
1954. Mississippian-Pennsylvanian boundary in northeastern Utah: Abstract. J. Paleontol., 28:506–507; also J. Sed. Petrology, 24:132.
1955. Carboniferous formations of northeastern Uinta Mountains. Wyoming Geol. Assoc. Field Conf. Guidebook. Pp. 49–59.
1957. Regional relations of Carboniferous rocks of northeastern Utah. Intermountain Assoc. Pet. Geol., 8th Ann. Field Conf. Guidebook. Pp. 56–77.
1959. Illustrated sections of strata adjacent to the Mississippian-Pennsylvanian boundary, western Uinta Mountains. Intermountain Assoc. Pet. Geol. 10th Ann. Field Conf. Guidebook. Pp. 75–81, 4 figs.
1960. Some preliminary aspects of Chainman stratigraphy. Intermountain Assoc. Pet. Geol. 11th Ann. Field Conf. Guidebook. Pp. 81–90, 2 figs.
1963. Visean-Namurian (Medial Chesterian) boundary in northern Utah and southeastern Idaho: Abstract. Geol. Soc. Amer. Spec. Paper 73. Pp. 94–95.

SCHAEFFER, F. E.
1960. Stratigraphy of the Silver Island Mountains. *In* Geology of the Silver Island Mountains. Utah Geol. Soc. Guidebook 15. Pp. 15–113.

SCOTT, A. J., S. P. ELLISON, JR., C. B. REXROAD, and W. ZIEGLER
1962. Comments on the orientation of conodonts. J. Paleontol., 36:1394–1396.

SPENCER, A. C.
1917. The geology and ore deposits of Ely, Nevada. U.S. Geol. Surv. Prof. Paper 96. 189 pp.

SPURR, J. E.
1903. Descriptive geology of Nevada south of the fortieth parallel and adjacent portions of California. U.S. Geol. Surv. Bull. 208. 229 pp.

STANLEY, E. A.
1958. Some Mississippian conodonts from the high resistivity shale of the Nancy Watson no. 1 well in northeastern Mississippi. J. Paleontol., 32(3):459–476, pls. 63–68, 2 text figs.

STAUFFER, C. R., and H. J. PLUMMER
1932 Texas Pennsylvania conodonts and their stratigraphic relations. Univ. Texas Bull. no. 3201, Pp. 13–50, pls. i-iv.

STURGEON, M. T., and W. YOUNGQUIST
1949. Allegheny conodonts from eastern Ohio. J. Paleontol., 23:380–386, pls. 74, 75.

THOMPSON, M. F.
1948. Studies of American fusulinids. Univ. Kans. Paleontol. Contrib., Protozoa, art. 1. 184 pp. 38 pls., 7 figs.

TSCHANZ, C. M., and E. H. PAMPEYAN
1961. Preliminary geologic map of Lincoln County, Nevada. U.S. Geol. Surv. Mineral Invest. Field Stud., Map MF-206.

VOGES, A.
1959. Conodonten aus dem Unterkarbon I und II (*Gattendorfia* und *Pericyclus*-stufe) des Sauerlandes. Palaontol. Z., 33:266–314, pls. 33–35.

WAGNER-GENTIS, C. H. T.
1963. Lower Namurian goniatites from the Griotte limestone of the Cantabric Mountain chain. Inst. Geol. y Min. Espana, Notas y Commun., no. 69. 23 pp., 8 pls.

WEBSTER, G. D., and N. G. LANE
1967. The Mississippian-Pennsylvanian boundary in southern Nevada. *In* Essays in Paleontology and Stratigraphy Raymond C. Moore commemorative volume. Univ. Kans. Press. Pp. 503–522, 4 figs.

WELLER, J. M., and others
1948. Correlation of the Mississippian formations of North America. Geol. Soc. Amer. Bull., 59:91–196, 1 chart.

WELLS, A. K., and J. R. KIRKALDY
1959. Outline of historical geology. George Allen & Unwin. 398 pp.

WELSH, J. E.
1959. Biostratigraphy of the Pennsylvanian and Permian systems in southern Nevada. Unpub. Ph.D. dissertation, Dept. Geol., Univ. Utah. 106 pp., 22 pls., 13 figs.

WESTGATE, L. G., and A. KNOPF
1932. Geology and ore deposits of the Pioche District, Nevada. U.S. Geol. Surv. Prof. Paper 171. 79 pp., 8 pls., 13 figs.

WILSON, E. C., R. H. WAINES, and A. H. COOGAN
1963. A new species of *Komia* Korde and the systematic position of the genus. Palaeontology, 6:246–253, pls. 34–35, 3 figs.

YOUNGQUIST, W., and R. H. DOWNS
1949. Additional conodonts from the Pennsylvanian of Iowa. J. Paleontol., 23:161–171, pls. 30–31.

YOUNQUIST, W., and B. C. HEEZEN
1948. Some Pennsylvanian conodonts from Iowa. J. Paleontol., 22:767–773, pl. 118.

YOUNGQUIST, W., and A. K. MILLER
1949. Conodonts from the Late Mississippian Pella beds of south-central Iowa. J. Paleontol., 23:617–622, pl. 101.

ZIEGLER, W.
1962. Die Conodonten aus den Geröllen Zechsteinkonglomerates von Rossenray (südwestlich Rheinberg/Niederrhein). Fortschritte in der Geologie von Rheinland und Westfalen, 6: 391–406, 4 pls., 1 fig.

PLATES

Magnifications given for fossils are approximate; all figures are ×40 unless otherwise indicated. Specimens were coated with ammonium chloride. Negatives of all photographs were made by the author and are on file in the Department of Geology, University of California, Los Angeles.

PLATE 2

(Plate 1 in pocket)

Fig. 1. View of Kane Springs Wash East (Section B), looking east. *Mc*, Chainman shale; *Mis*, Indian Springs formation; Mbs_1, Bird Spring formation unit 1; $¶bs_2$, $¶bs_3$, Bird Spring formation units 2 and 3, respectively.

Fig. 2. View of lower part of section at Arrow Canyon (Section D), looking northeast. *Mbw*, Battleship Wash formation; *Mis*, Indian Springs formation; Mbs_1, Bird Spring formation unit 1; $¶bs_2$, Bird Spring formation unit 2.

PPbs3
PPbs2
Mbs1
Mis
Mc
1

PPbs2
Mbs1
Mis
Mbw
2

PLATE 3

Fig. 1. View of Bird Spring formation, upper beds of unit 2 in Arrow Canyon (Section D), looking northeast.

Fig. 2. View of bench-slope forming beds in the lower part of unit 4 of the Bird Spring formation in Las Vegas range (Section F), looking northeast. Arrow Canyon range in distance.

PLATE 4

Figs. 1, 4, 5. *Cavusgnathus spathus* Dunn. Figs. 1a, 1b, oral and outer lateral views, respectively; hypotype no. 43463. Fig. 4 oral view; hypotype no. 43462. Figs. 5a, 5b, oral and inner lateral views, respectively; hypotype no. 43461. Page 28.

Fig. 2. *Cavusgnathus altus* Harris and Hollingsworth. Figs. 2a, 2b, oral and inner lateral views, respectively; hypotype no. 43457. Page 25.

Fig. 3. *Cavusgnathus naviculus* (Hinde), oral view; hypotype no. 43465. Page 28.

Fig. 6. *Cavusgnathus gigantus* Gunnell. Figs. 6a, 6b, oral and inner lateral views, respectively; hypotype no. 43458. Page 26.

Figs. 7, 8. *Streptognathodus noduliferus* (Ellison and Graves). Figs. 7a, 7b, 7c, oral, aboral, and inner lateral views, respectively; hypotype no. 43488. Figs. 8a, 8b, 8c, oral, inner lateral, and aboral views, respectively; hypotype no. 43489. Page 48.

Fig. 9. *Cavusgnathus lautus* Gunnell. Figs. 9a, 9b, oral and inner lateral views, respectively; hypotype no. 43459. Page 28.

Fig. 10. *Cavusgnathus convexus* Rexroad. Figs. 10a, 10b, inner lateral and oral views, respectively; hypotype no. 43456. Page 26.

Fig. 11. *Streptognathodus suberectus* Dunn, oral view; hypotype no. 43484. Page 49.

Fig. 12. *Magnilaterella robusta* Rexroad and Collinson. Figs. 12a, 12b, anterior and posterior views, respectively; hypotype no. 43454. Page 38.

Fig. 13. *Streptognathodus unicornis* Rexroad and Burton. Figs. 13a, 13b, oral and inner lateral views, respectively; hypotype no. 43495. Page 49.

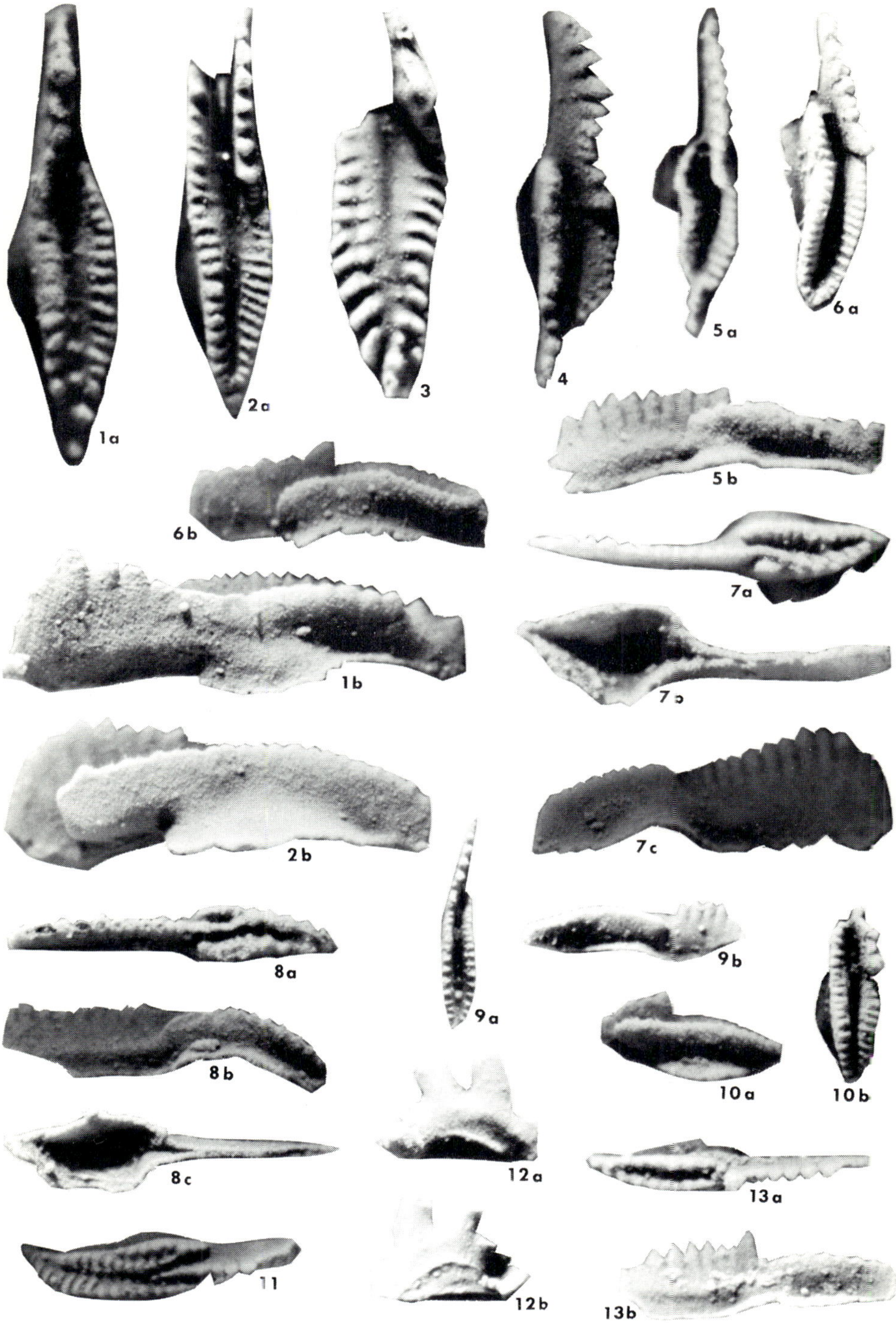
1a
2a
3
4
5a
6a
5b
6b
7a
1b
7b
2b
7c
8a
9b
9a
10a
10b
8b
12a
8c
13a
11
12b
13b

PLATE 5

Figs. 1–7. *Gnathodus muricatus* (Dunn). Figs. 1a, 1b, inner lateral and oral views, respectively, of a gerontic individual (specimen was broken across posterior part of blade during preparation); hypotype no. 43470. Figs. 2a 2b, oral and outer lateral views, respectively; hypotype no. 43473. Fig. 3, oral view; hypotype no. 43476. Fig. 4, oral view; hypotype no. 43475. Fig. 5, oral view; hypotype no. 43474. Fig. 6, oral view; hypotype no. 43471. Fig. 7, oral view; hypotype no. 43472. Page 32.

Fig. 8. *Gnathodus girtyi simplex* Dunn transitional to *Gnathodus muricata* (Dunn), oral view; hypotype no. 43477. Page 33.

Figs. 9, 14, 15. *Gnathodus bassleri* (Harris and Hollingsworth). Figs. 9a, 9b, 9c, oral, inner lateral and aboral views, respectively; hypotype no. 43478. Fig. 14, oral view; hypotype no. 43479. Figs. 15a, 15b, oral and outer lateral views, respectively; hypotype no. 43480. Page 29.

Fig. 10. *Gnathodus girtyi simplex* Dunn, oral view; hypotype no. 43469. Page 32.

Figs. 11, 12. *Gnathodus bilineatus* (Roundy). Fig. 11, oral view; hypotype no. 43467. Fig. 12, oral view; hypotype no. 43466. Page 30.

Fig. 13. *Gnathodus commutatus* Branson and Mehl. Figs. 13a, 13b, oral and inner lateral views, respectively; hypotype no. 43468. Page 31.

Fig. 16. *Gnathodus defectus* Dunn, oral view; hypotype no. 43483. Page 32.

Figs. 17, 18. *Idiognathoides convexus* (Ellison and Graves). Fig. 17, oral view; hypotype no. 43498. Figs. 18a, 18b, oral and inner lateral views respectively; hypotype no. 43496. Page 37.

Figs. 19–21. *Gnathodus opimus* Igo and Koike. Fig. 19, oral view; hypotype no. 43482. Fig. 20, oral view; hypotype no. 43481. Figs. 21a, 21b, 21c, aboral, oral and inner lateral views, respectively; hypotype no. 43497. Page 33.

Fig. 22. *Hibbardella ortha* Rexroad, posterior view; hypotype no. 43529. Page 34.

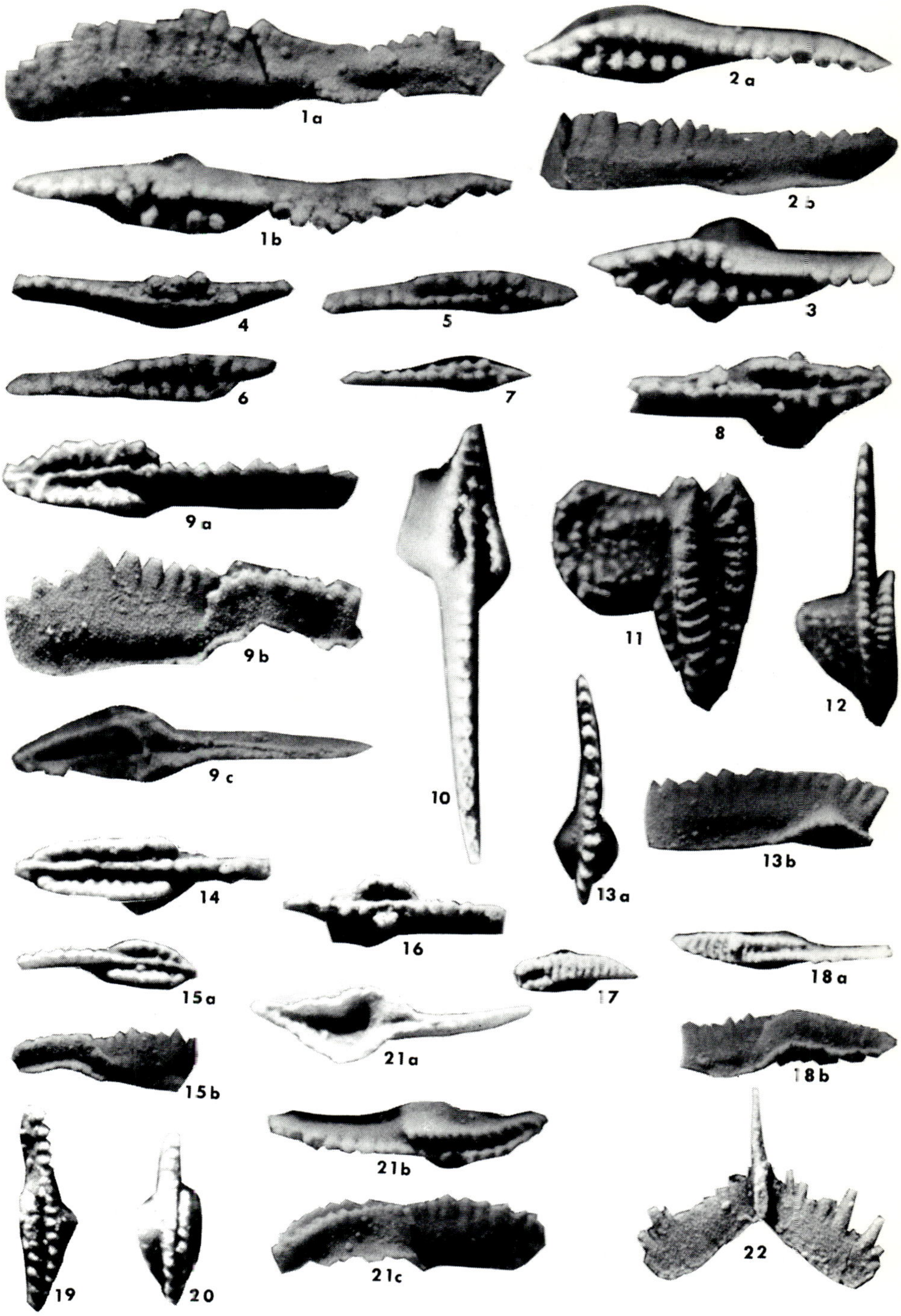
1a
2a
2b
1b
4
5
3
6
7
8
9a
10
11
12
9b
9c
13a
13b
14
16
15a
17
18a
21a
15b
18b
21b
19
20
21c
22

PLATE 6

Figs. 1–5. *Streptognathodus expansus* Igo and Koike. Fig. 1, oral view; hypotype no. 43494. Figs. 2a, 2b, oral and outer lateral views, respectively; hypotype no. 43492. Figs. 3a, 3b, oral and outer lateral views, respectively; hypotype no. 43490. Fig. 4, oral view; hypotype no. 43491. Fig. 5, oral view; hypotype no. 43493. Page 46.

Figs. 6–12. *Idiognathodus delicatus* Gunnell. Fig. 6, oral view; hypotype no. 43502. Fig. 7, oral view; hypotype no. 43501. Fig. 8, oral view; hypotype no. 43504. Figs. 9a, 9b, oral and inner lateral views, respectively; hypotype no. 43506. Fig. 10, oral view; hypotype no. 43505. Fig. 11, oral view; hypotype no. 43500. Fig. 12, oral view; hypotype no. 43503. Page 35.

Fig. 13. *Idiognathodus claviformis* Gunnell. Figs. 13a, 13b, 13c, oral, inner lateral, and aboral views, respectively; hypotype no. 43499. Page 35.

Figs. 14, 15. *Streptognathodus lanceolatus* n. sp. Figs. 14a, 14b, 14c, oral, outer lateral, and aboral views, respectively; paratype no. 43486. Figs. 15a, 15b, 15c, oral, outer lateral, and aboral views, respectively; holotype no. 43485. Page 47.

Figs. 16–20. *Hindeodella* sp. Fig. 16, inner lateral view; hypotype no. 43546. Fig. 17, inner lateral view; hypotype no. 43544. Fig. 18, lateral view; hypotype no. 43545. Fig. 19, inner lateral view; hypotype no. 43542. Fig. 20, lateral view; hypotype no. 43543. Page 35.

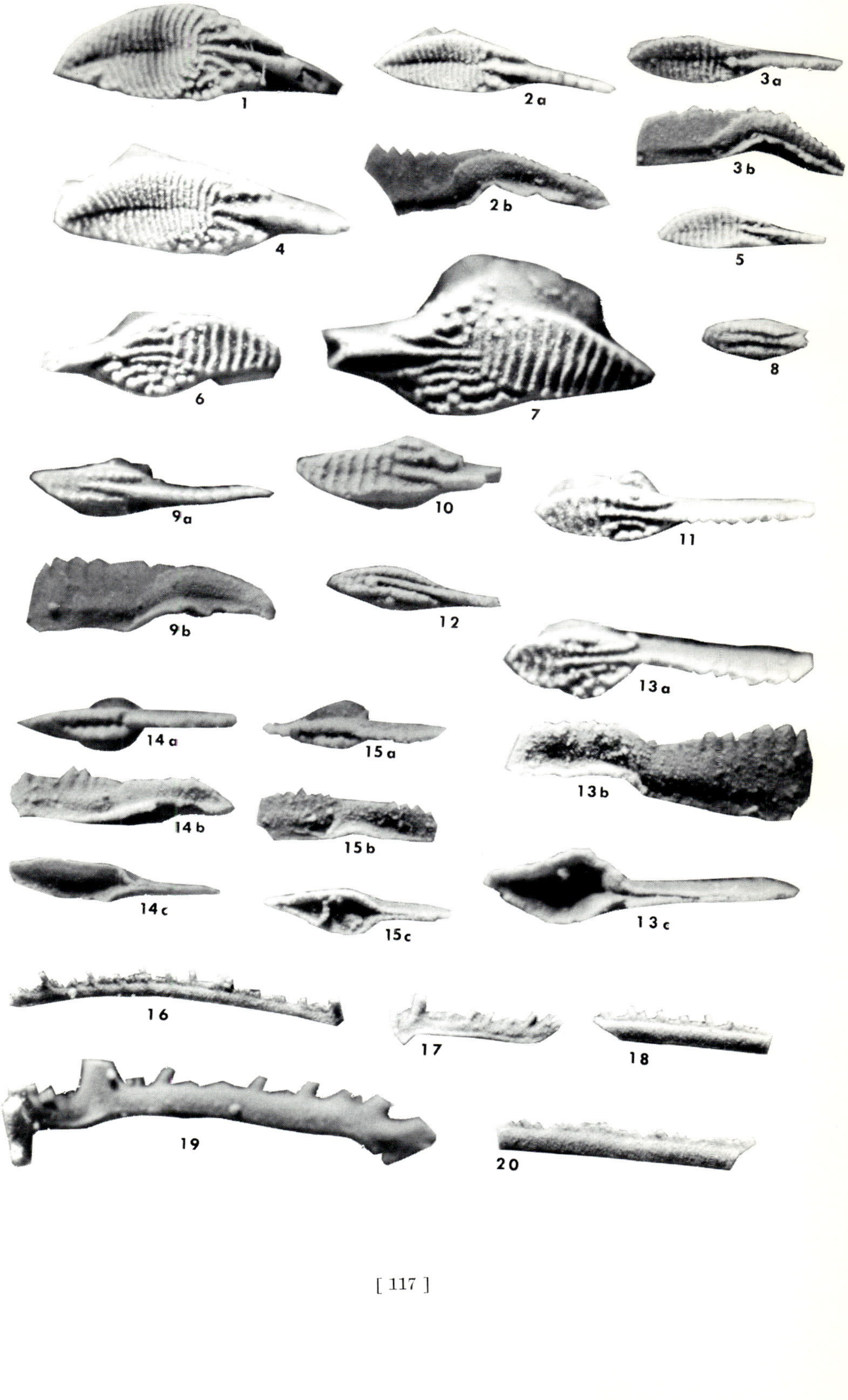
1
2a
3a
3b
2b
4
5
6
7
8
9a
10
11
9b
12
13a
14a
15a
13b
14b
15b
14c
15c
13c
16
17
18
19
20

PLATE 7

Figs. 1–3. *Spathognathodus rexroadi* n. sp. Figs. 1a–1d, inner lateral, oral, outer lateral, and aboral views, respectively; holotype no. 43510. Fig. 2, outer lateral view; paratype no. 43513. Fig. 3, inner lateral view; paratype no. 43511. Page 45.

Fig. 4. *Spathognathodus minutus* (Ellison), inner lateral view; hypotype no. 43509. Page 44.

Fig. 5. *Spathognathodus campbelli* Rexroad, inner lateral view; hypotype no. 43507. Page 43.

Fig. 6. *Spathognathodus* sp., inner lateral view; hypotype no. 43515. Page 45.

Fig. 7. *Spathognathodus coloradoensis* Murray and Chronic, inner lateral view; hypotype no. 43508. Page 44.

Fig. 8. *Ozarkodina roundyi* (Hass), inner lateral view; hypotype no. 43526. Page 43.

Fig. 9. *Ozarkodina campbelli* Youngquist and Heezen, inner lateral view; hypotype no. 43523. Page 41.

Fig. 10. *Ozarkodina curvata* Rexroad, outer lateral view; hypotype no. 43524. Page 42.

Fig. 11. *Ozarkodina delicatula* (Stauffer and Plummer), inner lateral view of incomplete specimen; hypotype no. 43525. Page 42.

Fig. 12. *Neoprioniodus loxus* Rexroad, inner lateral view; hypotype no. 43517. Page 33.

Fig. 13. *Neoprioniodus scitulus* (Branson and Mehl), inner lateral view; hypotype no. 43518. Page 39.

Fig. 14. *Neoprioniodus singularis* (Hass), outer lateral view; hypotype no. 43519. Page 40.

Figs. 15, 16. *Neoprioniodus? expandofundus* n. sp. Figs. 15a–15d, outer lateral, inner lateral, aboral, and oral views, respectively; holotype no. 43521. Figs. 16a–16c, oral, outer lateral, and aboral views, respectively; paratype no. 43522. Page 41.

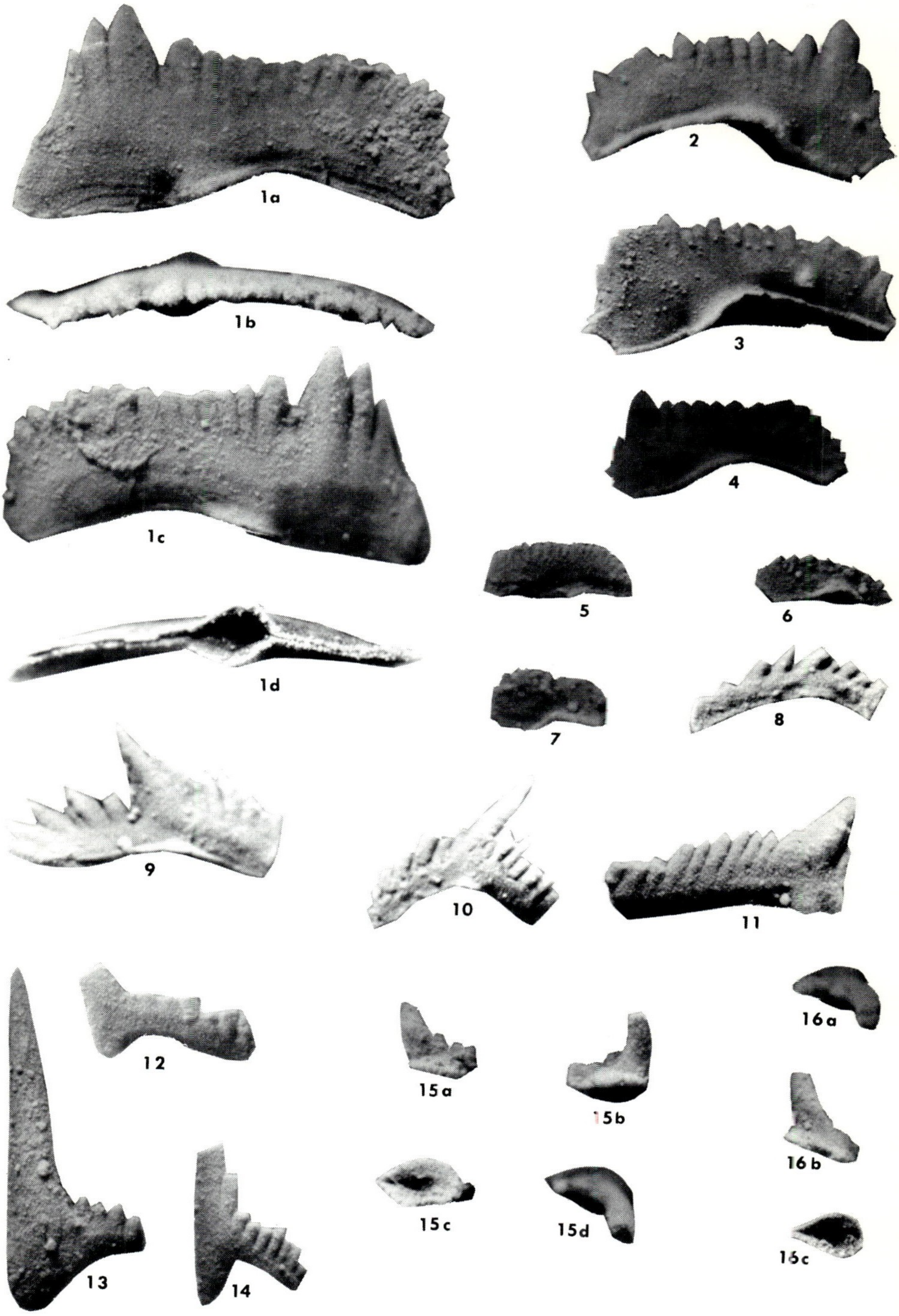
1a
2
1b
3
1c
4
5
6
1d
7
8
9
10
11
12
15a
16a
15b
16b
15c
15d
13
14
16c

PLATE 8

(Figure 14 ×30)

Fig. 1. *Geniculatus?* sp. Figs. 1a, 1b, inner lateral and oral views, respectively; hypotype no. 43455. Page 29.

Figs. 2–5. *Lonchodina* sp. Fig. 2, inner lateral view; hypotype no. 43539. Fig. 3, inner lateral view; hypotype no. 43538. Fig. 4, inner lateral view; hypotype no. 43541. Fig. 5, inner lateral view; hypotype no. 43540. Page 38.

Fig. 6. *Metalonchodina* sp., posterior view; hypotype no. 43533. Page 39.

Figs. 7, 8. *Roundya* sp. Figs. 7a, 7b, lateral and posterior views, respectively; hypotype no. 43531. Figs 8a, 8b, posterior and lateral views, respectively; hypotype no. 43532. Page 43.

Figs. 9–12. *Ligonodina* sp. Fig. 9, inner lateral view; hypotype no. 43534. Fig. 10, posterior view; hypotype no. 43536. Fig. 11, inner lateral view; hypotype no. 43537. Fig. 12, inner lateral view; hypotype no. 43535. Page 38.

Fig. 13. *Neoprioniodus* cf. *N. camurus* Rexroad. Figs. 13a, 13b, inner lateral and outer lateral views, respectively; hypotype no. 43516. Page 39.

Fig. 14. *Hibbardella* sp. A, lateral view; hypotype no. 43530. Page 34.

Fig. 15. *Synprioniodina microdenta* Ellison, inner lateral view; hypotype no. 43528. Page 50.

Fig. 16. *Synprioniodina denticamura* Rexroad and Liebe. Figs. 16a, 16b, inner lateral and outer lateral views, respectively; hypotype no. 43527. Page 50.

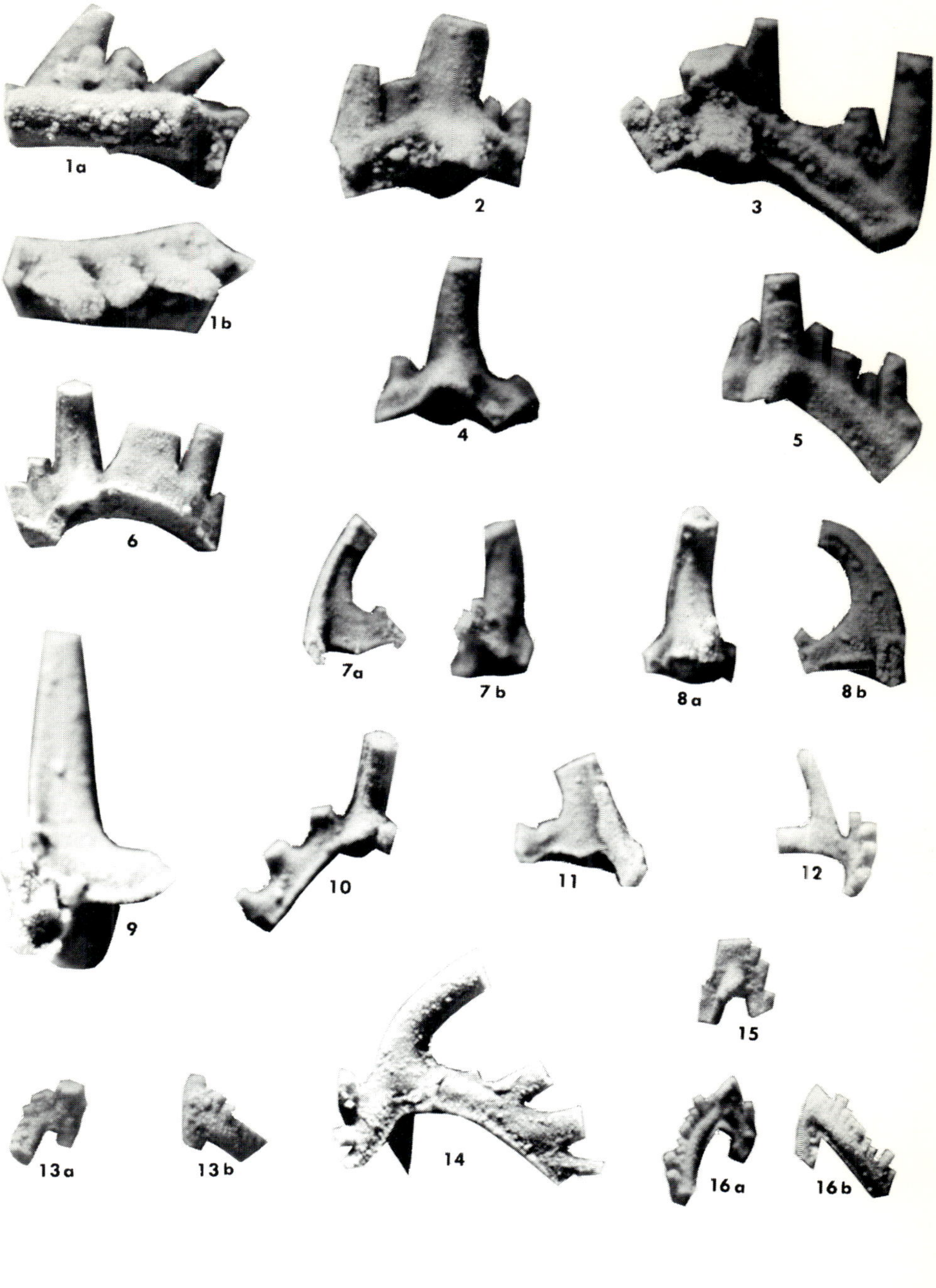
1a
2
3
1b
4
5
6
7a
7b
8a
8b
9
10
11
12
15
13a
13b
14
16a
16b